Functional Composites

Scrivener Publishing
100 Cummings Center, Suite 541J
Beverly, MA 01915-6106

Advances in Production Engineering

Series Editor: Dr. Sandip Kunar (sandip.sandip.kunar@gmail.com)

"Advances in Production Engineering" addresses recent developments and research issues related to the production engineering which necessitates the development of new materials and manufacturing processes in order to produce high-quality products at lower prices and in less time. This series aims to provide a scientific platform for researchers, practitioners, professionals, and academicians to discuss the most recent technological developments in metals, polymers, ceramics, composites, biomaterials, nanomaterials, special materials, metals, micro-forming, powder metallurgy, ceramics processing, non-traditional machining, high speed machining, micro and nanomachining, and laser processing. Tribological analysis, friction behavior, modelling, and optimization techniques in materials, machining, and manufacturing are also covered in the series.

Publishers at Scrivener
Martin Scrivener (martin@scrivenerpublishing.com)
Phillip Carmical (pcarmical@scrivenerpublishing.com)

Functional Composites

Role in Modern Engineering

Edited by

Sandip Kunar

Department of Mechanical Engineering, Aditya University, Surampalem, Andhra Pradesh, India

Pranav Charkha

School of Engineering & Technology, Pimpri Chinchwad University, Pune, Maharashtra, India

Santosh Jaju

Department of Mechanical Engineering, G. H. Raisoni College of Engineering, Nagpur, Maharashtra, India

and

Harish Tiwari

Pimpri Chinchwad College of Engineering & Research, Ravet, Pune, India

Scrivener Publishing

WILEY

This edition first published 2025 by John Wiley & Sons, Inc., 111 River Street, Hoboken, NJ 07030, USA and Scrivener Publishing LLC, 100 Cummings Center, Suite 541J, Beverly, MA 01915, USA

For more information about Scrivener publications please visit www.scrivenerpublishing.com.

Wiley Global Headquarters
111 River Street, Hoboken, NJ 07030, USA

For details of our global editorial offices, customer services, and more information about Wiley products visit us at www.wiley.com.

The manufacturer's authorized representative according to the EU General Product Safety Regulation is Wiley-VCH GmbH, Boschstr. 12, 69469 Weinheim, Germany, e-mail: Product_Safety@wiley.com.

Library of Congress Cataloging-in-Publication Data

ISBN 978-1-394-24200-9

Cover image: Generated with AI using Adobe Firefly
Cover design by Russell Richardson

Set in size of 11pt and Minion Pro by Manila Typesetting Company, Makati, Philippines

Printed in the USA

10 9 8 7 6 5 4 3 2 1

Contents

Preface **xv**

1 Introduction to Functional Composite Materials **1**
Sandip Kunar, Gurudas Mandal, Jagadeesha T. and Pranav Charkha
1.1 Introduction 1
1.2 Overview 3
1.3 Characteristics of Composites 7
1.4 A Fundamental Method for Choosing Materials 8
1.5 Polymer Matrix 9
1.6 Reinforcements 11
1.7 Techniques for Producing Composites 12
1.7.1 Molding Using Open Contact 13
1.7.2 Resin Infusion Method 13
1.7.3 Injection Molding 14
1.7.4 Filament Winding 14
1.7.5 Pultrusion Process 14
1.7.6 Additive Manufacturing 15
1.8 Composite Properties 15
1.9 Latest Developments 16
1.10 Applications 17
1.11 Conclusion 17
References 17

2 Shape Memory Alloys as Functional Composites **21**
Param Singh, Yatin Khanna, Roopak Varshney and Yajush Walia
2.1 Introduction 22
2.1.1 Exploration and Attempts at Development 22

2.2 Composition and Microstructure of Shape Memory Alloys' (SMAs') Composition 23
2.2.1 Microstructure 24
2.2.2 Characterization Techniques 26
2.2.3 Shape Memory Effect 26
2.2.4 Super Elasticity 28
2.3 Processing Techniques Used for Shape Memory Alloys 30
2.3.1 Casting 30
2.3.2 Powder Metallurgy 31
2.3.3 Thermomechanical Processing 32
2.4 Characterization Methods Employed to Evaluate the Microstructural and Mechanical Properties of SMAs 33
2.4.1 Microscopy 33
2.4.2 X-Ray Diffraction (XRD) 34
2.4.3 Tests of the Material's Mechanical Characteristics and Behavior 35
2.5 Applications of Shape Memory Alloys as Functional Composites 35
2.5.1 Aerospace 36
2.5.2 Automotive 36
2.5.3 Medical Applications 36
2.5.4 Consumer Electronics 37
2.6 Design Considerations and Challenges in Using SMAs for Specific Applications 38
2.6.1 Compatibility with the Prerequisites of the Application 38
2.6.2 Mechanisms of Actuation and Their Controls 38
2.7 The Choice of Materials and Their Compatibility 39
2.7.1 Fatigue and Durability 39
2.7.2 Manufacturing and Processing 40
2.7.3 Expense and the Possibility of Commercialization 40
2.8 Case Studies and Success Stories That Demonstrate the Practical Implementation of SMAs as Functional Composite Structures 41
2.8.1 The Use of Smart Morphing Adaptors in Aerospace 41
2.8.2 Self-Deployable Space Structures 41
2.8.3 Orthopedic Implants and Medical Devices 41
2.8.4 Adaptive Structures in the Automotive Sector 42
2.8.5 Wearable Technologies and Intelligent Textiles 42
2.9 Current State of Research in Shape Memory Alloys and Potential Areas for Future Exploration 42

2.10 Recent Advancements in the Development of Novel SMA Compositions, Processing, and Applications 44
2.10.1 Novel SMA Compositions 44
2.10.2 Advanced Processing Methods 45
2.10.3 SMA Application Advancements 45
2.10.4 Composite Hybrid Structures 46
2.11 Conclusions 46
References 47

3 Characterization and Testing of Smart Functional Composites 49
Ranjita Swain, Sunita Routray and Rudra Narayan Mohapatro
3.1 Introduction 50
3.1.1 Shape Memory Composites (SMPs) 50
3.1.2 Self-Healing Composites 50
3.1.3 Piezoelectric Composites 51
3.1.4 Magnetostrictive Composites 51
3.1.5 Thermoelectric Composites 51
3.1.6 Conductive Composites 52
3.1.7 Light-Responsive Composites 52
3.1.8 Bio-Inspired Composites 52
3.1.9 Multi-Functional Composites 53
3.2 Mechanical Characterization 54
3.2.1 Overview of Mechanical Testing Methods 54
3.2.2 Discussion of Mechanical Properties 56
3.2.3 Mechanical Characterization of Smart Functional Composites 57
3.2.4 Electrical Characterization of Smart Functional Composites 58
3.2.4.1 Dielectric Strength 58
3.2.4.2 Insulation Resistance 58
3.2.4.3 Volume Resistivity and Surface Resistivity 58
3.2.5 Types of Smart Functional Composites 59
3.2.5.1 Structural Smart Composites 59
3.2.5.2 Composites for Actuation 59
3.2.5.3 Novel Functional Composites 60
3.2.5.4 Nanocomposites for Novel Functions 60
3.3 Thermal Characterization of Smart Functional Composites 60
3.3.1 Thermal Behavior of Smart Functional Composites 61
3.3.1.1 Thermal Characterization Techniques 62

3.3.1.2 Thermal Properties of Key Smart Functional Composites 62
3.3.1.3 Challenges and Future Directions 63
3.4 Environmental and Durability Testing of Functional Smart Materials 63
3.4.1 Thermal Cycling and Temperature Testing 64
3.4.2 Moisture and Humidity Testing 64
3.4.3 UV and Radiation Exposure 65
3.4.4 Mechanical and Vibration Testing 65
3.4.5 Electromagnetic Interference (EMI) Testing 65
3.5 Durability Testing Methodologies for Smart Functional Composites 66
3.5.1 Accelerated Aging Testing 66
3.5.2 Self-Healing and Damage Detection 66
3.5.3 Field Testing and Real-World Simulations 67
3.6 Recent Advances in Smart Functional Composite Testing 67
3.6.1 Integration of Smart Sensors for Real-Time Monitoring 67
3.6.2 Nanomaterial Enhancements 67
3.7 Conclusion 67
References 68

4 Piezoelectric Nanocomposites 71
Param Singh, Srijal Mishra, Roopak Varshney and Yajush Walia
4.1 Introduction 72
4.1.1 The Piezoelectric Effect 72
4.1.1.1 Direct Piezoelectric Effect 73
4.1.1.2 Reverse Piezoelectric Effect 73
4.2 Variables and Constants That Have an Impact on the Performance of Piezoelectric Materials 74
4.2.1 Electro-Mechanical Coupling Factors (k) 74
4.2.2 Piezoelectric Strain (Charge) Constant (d) 75
4.2.3 Piezoelectric Voltage Constants (g) 76
4.2.4 Mechanical Quality Factor (Qm) 77
4.2.5 Electrical Loss ($\tan\delta$) 77
4.2.6 Dielectric Constant (ε) 77
4.3 Piezoelectric Nanocomposites 78
4.3.1 Piezoelectric Nanocomposite Materials—Polymer-Based 79
4.3.2 Poling 81

4.3.3 Preparation of a Nano-Polymeric Piezoelectric Composite 83
4.3.4 Piezoelectric Nanoparticle Polymer Composite Foam (PNPF) 84
4.3.5 Reverse Effect of PNPF 84
4.4 Piezoelectric Polymer Materials 85
4.4.1 Polyvinylidene Fluoride (PVDF) 85
4.4.2 Polyvinylidene Fluoride Trifluoro Ethylene (PVDF-TrFE) 86
4.4.3 Polyvinylidene Cyanide-Vinyl Acetate 86
4.4.4 Polyamide 11 (Nylon 11) 86
4.4.5 Cellular Polypropylene (PP) 86
4.4.6 Poly-Organo-Phosphazenes (POPh) 87
4.5 Piezoelectric Nanocomposite Materials—Ceramic-Based 87
4.5.1 Lead Zirconate Titanate (PZT) 87
4.5.2 Potassium Sodium Niobate (KNN) 88
4.5.3 Bismuth Sodium Titanate (BNT) 88
4.5.4 Aluminum Nitride (AlN) 88
4.5.5 Lithium Niobate 89
4.6 Improvements to Piezoelectric Ceramics 89
4.7 Applications of Piezoelectric Nanocomposites 89
4.7.1 Bio-Medical 89
4.7.2 Piezoelectric Tactile Sensors 89
4.7.3 Piezoelectric Vibrational Energy Harvestors Using Polymer Nanocomposite 90
4.7.4 Piezoelectric Nano-Generator (PENG) 90
4.7.4.1 Force Applied Perpendicular to the Nanowire's Axis 91
4.7.4.2 Force Applied Parallel to the Nanowire's Axis 92
4.7.5 Nanocomposite Electrical Generators (NEG) 93
References 94

5 Modulation of Waveform Effect on Ni/Nano-ZrO_2–TiO_2 Composite Coating on Mild Steel 99
Koona Bhavani, VSN Venkata Ramana, Chitrada Prasad, A. Ramesh and Marana Lalitha
5.1 Introduction 100
5.2 Procedure 101
5.2.1 Materials and Synthesis of Nanocomposites 101
5.2.2 Electrodeposition 102
5.2.3 Characterization 104

5.3 Results and Discussions 104
5.3.1 Microscopic Structure Analysis 104
5.3.2 X-Ray Diffraction 106
5.3.3 Microhardness 108
5.3.4 Pitting Corrosion Studies 109
5.4 Conclusions 111
References 111

6 Smart Composite Materials for Aerospace Applications 115
Ibrahim Momoh-Bello Omiogbemi, Emmanuel Imhanote Awode, Mohammed Habib Muhammad, Adegoke Adesanmi and Ishaya Musa Dagwa
6.1 Introduction 116
6.1.1 Importance of Composite Materials in Aerospace 118
6.1.1.1 Weight Reduction 119
6.1.1.2 High Strength-to-Weight Ratio 119
6.1.1.3 Corrosion Resistance 119
6.1.1.4 Design Flexibility 120
6.1.1.5 Heat Resistance and Thermal Stability 120
6.1.1.6 Structural Health Monitoring 121
6.1.1.7 Reduction in Production Costs 121
6.1.1.8 Sustainability 122
6.1.2 Overview of Smart Functional Composite Polymer Materials 122
6.1.2.1 Key Characteristics 122
6.1.2.2 Manufacturing Techniques 123
6.1.2.3 Applications 124
6.2 Types of Smart Functional Composite Polymer Materials 124
6.2.1 Polymer Matrix Composites (PMCs) 125
6.2.2 Carbon Fiber Reinforced Polymers (CFRP) 125
6.2.3 Glass Fiber Reinforced Polymers (GFRP) 126
6.2.4 Shape Memory Polymers (SMPs) 127
6.2.5 Self-Healing Polymers (SHPs) 129
6.2.6 Conductive Polymer Composites (CPCs) 130
6.3 Properties and Characteristics 131
6.3.1 Mechanical Properties 131
6.3.1.1 Strength 132
6.3.1.2 Stiffness 132
6.3.1.3 Toughness 133

6.3.2 Thermal Properties 133
6.3.2.1 Thermal Conductivity 133
6.3.2.2 Thermal Stability 135
6.3.3 Electrical Properties 136
6.3.3.1 Conductivity 136
6.3.3.2 Resistivity 136
6.3.4 Multifunctional Properties 137
6.3.4.1 Self-Sensing 137
6.3.4.2 Self-Healing 137
6.4 Aerospace Applications 138
6.4.1 Structural Components 138
6.4.1.1 Wings 138
6.4.1.2 Fuselage 139
6.4.1.3 Control Surfaces 139
6.4.2 Functional Components 139
6.4.2.1 Actuators 139
6.4.2.2 Sensors 139
6.4.2.3 Energy 140
6.4.2.4 Storage 140
6.4.3 Thermal Management Systems 140
6.4.4 Self-Healing Coatings and Materials 141
6.5 Manufacturing Techniques 141
6.5.1 Conventional Methods 142
6.5.1.1 Hand Layup 142
6.5.1.2 Resin Transfer Molding (RTM) 142
6.5.1.3 Compression Molding 143
6.5.2 Advanced Methods 144
6.5.2.1 Additive Manufacturing (AM) 144
6.5.2.2 Electrospinning 145
6.5.2.3 Automated Fiber Placement (AFP) 145
6.5.3 Nanotechnology-Enabled Manufacturing 146
6.6 Challenges 147
6.6.1 Interfacial Properties and Bonding 147
6.6.2 Scalability and Cost-Effectiveness 148
6.6.3 Integration with Existing Aerospace Systems 148
6.6.4 Future Research Directions 148
6.7 Conclusion 149
References 150

7 **Behavioral Study of Tribological Coating of Smart Functional Composites for High Wear Applications** **169**
M. Sasi Kumar, N. Venkatesh, M. Makesh Kumar, S.L. Pradeep Kumar, D. Santhosh Kumar and B. Deeban
7.1 Introduction 170
7.2 Principle of Tribology and Mechanisms of Wear 171
7.2.1 Wear Types 171
7.2.2 Significance of Coating in Enhancing Wear Resistance 172
7.3 Intelligent Multifunctional Composites 172
7.3.1 Categories of Functional Composites 173
7.3.2 Applications in High-Wear Environments 173
7.4 Tribological Coatings 173
7.5 Techniques for the Design and Fabrication of Materials 175
7.6 Development of Performance 176
7.7 Obstacles and Constraints 179
7.7.1 Material Compatibility Challenges 179
7.7.2 Considerations Regarding Expenses and the Ability to Scale 179
7.7.3 Constraints Related to Environmental and Operational Factors 179
7.8 Emerging Trends and Research Pathways 180
7.9 Conclusion 182
References 183

8 **Applications to Dynamic Behavior: Free Vibration Analysis of Functional Composites** **189**
Debarupam Gogoi, Pramod Kumar Parida, Md. Irquam Alam, Mihir Kumar Pandit and Arun Kumar Pradhan
8.1 Introduction 190
8.1.1 Advantages of FGM Over Traditional Composites 191
8.2 Recent Advances in the Development and Application of Functionally Graded Materials 194
8.3 Challenges and Future Directions 196
8.4 Free Vibration Analysis of Functional Composite Materials 197
8.4.1 Geometry and Property 197
8.4.2 Mathematical Modeling 200
8.5 Conclusion 210
References 211

9 Structural Health Monitoring of Composite Structures: Utility, Challenges, Sensor Technologies, and Advanced Methods **215**
Sumanta Banerjee and Anindita Kundu
9.1 Introduction 216
9.1.1 Traditional Composites: Overview and Advantages 216
9.1.2 Smart Composites: Characterization and Application Potential 218
9.1.3 Structural Health Monitoring: Why the Need? 219
9.1.4 What This Chapter is All About? 219
9.2 Composite Materials: Problems and Challenges 220
9.3 Structural Health Monitoring for Composites: Benefits and Challenges 221
9.4 Sensors Employed in SHM: A Review 222
9.4.1 Fiber Optic Sensors (FOSs) 222
9.4.2 Resistance Strain Gauges 224
9.4.3 Piezoelectric Sensors 225
9.4.4 Eddy Current Sensors 225
9.4.5 Micro-Electromechanical (MEMS) Sensors 226
9.5 Overview of Typical SHM Methods for Composite Materials 229
9.5.1 Lamb Wave Method (LWM)-Based SHM 229
9.5.2 The Electromechanical (E/M) Impedance Method 232
9.5.3 Active Vibration-Based Method (AVBM) 234
9.5.4 Acoustic Emission "Passive" Methods for SHM 236
9.5.5 Strain-Based Methods for SHM 238
9.5.6 The Comparative Vacuum Monitoring (CVM) "Passive" Method 240
9.6 Conclusions and Future Scope 240
References 241

Index **247**

Preface

Continuous fiber-reinforced composites is a novel class of materials with characteristics such as high modulus, tolerance, strength, damage resistance, and safety. The aerospace industry now relies heavily on them, and they are increasingly used in consumer goods, recreational products, and other industries. Polymer matrix composites are widely used across many industries. In the field of materials science and engineering, they represent a vital subject for undergraduate students. Due to the dominance of traditional composite materials in structural applications (such as aerospace), their use has largely focused on mechanical and fabrication characteristics. However, driven by the demands of the battery, biomedical, and electronic industries, non-structural applications are becoming increasingly significant. Since functional and structural composites have different performance and cost requirements, the scientific principles guiding their design also differ considerably. This book's focus on functional composite materials sets it apart from other related works.

In addition to discussing the creation, composition, characteristics, and applications of relevant composite materials, the book presents the fundamental concepts underlying each function. While cement-matrix composites are commonly used for structural purposes, ceramic and metal-matrix composites are becoming increasingly important. This book covers composite materials across all matrix types.

Classifying composite materials by their matrix type is a less common approach in books on composites. In contrast, this book adopts a novel functional approach, organizing materials according to their applications. This helps readers understand how composites are designed to meet the needs of various industries. Such insight benefits both professionals across sectors and students preparing for industry roles. Moreover, the functional approach allows for the systematic exploration of a broad range of scientific topics beyond just mechanical behavior.

The book contains nine chapters. Chapter 1 provides an overview of composite materials, covering the properties of polymer matrices and reinforcements, material selection basics, production methods, composite properties, and applications. Chapter 2 focuses on the synthesis and processing of Shape Memory Alloys (SMAs) within composites. It addresses scalability, manufacturing techniques, performance optimization, and the challenges and potential of SMA-based composites. The chapter highlights the transformative potential of SMAs in driving advanced technological developments and offers a comprehensive analysis of their interaction with composite materials, including insights into their synthesis, applications, and future prospects.

Chapter 3 introduces key types of smart composites, including those embedded with piezoelectric, thermochromic, and magnetostrictive elements, highlighting their unique ability to enhance structural performance. It also explores challenges in their development, such as material compatibility, processing methods, and long-term durability. The chapter concludes by identifying emerging trends, such as the use of sustainable materials and the integration of artificial intelligence (AI) for real-time monitoring and adaptive response.

Chapter 4 covers the fundamentals of piezoelectric materials, offering a detailed discussion on piezoelectric nanocomposites, their properties, and the fabrication of select examples. These materials have broad applications, including nano-transformers, portable nano-generators, energy harvesters, and biomedical devices—though most are used in sensor manufacturing. Chapter 5 explores the electrodeposition process using periodic modulation of rectangular and triangular waveforms, examining their effect on the conventional electroplating of Ni/ZrO_2-TiO_2 coatings on mild steel.

Chapter 6 explores the evolving field of smart functional composite polymer materials for aerospace applications. It examines their use in structural components, functional devices, thermal management systems, and self-healing coatings. Chapter 7 focuses on the friction and wear resistance behavior of advanced coatings, highlighting their ability to enhance mechanical performance under extreme conditions. The chapter illustrates how these coatings can extend component lifespan, reduce maintenance needs, and offer sustainable engineering solutions for high-demand sectors such as aerospace, automotive, and energy.

Chapter 8 discusses recent developments in composite materials, highlighting their advantages and limitations, trends in advanced and functionally graded materials (FGMs), various FGM structures and their applications, and the dynamic behavior of FGM plate structures. Chapter 9 provides a comprehensive review of structural health monitoring (SHM) techniques for composites, outlining both their benefits and implementation challenges. It examines various sensor technologies—including Fiber Optic Sensors (FOSs), Resistance Strain Gauges, Eddy Current Sensors, Piezoelectric Sensors, and MEMS Sensors—for their applicability in SHM. The chapter also categorizes key SHM methodologies into 'active' and 'passive' approaches, covering techniques such as Lamb waves, electro-mechanical impedance, active vibration monitoring, strain-based methods, acoustic emission, and Comparative Vacuum Monitoring (CVM).

This book can serve as a reference for undergraduate and postgraduate students, as well as engineers, technicians, technology managers, and marketing staff. It will be especially helpful for thermal, electrical, industrial, and chemical engineers, thanks to the broad scientific scope enabled by the functional approach. Finally, we extend our gratitude to Martin Scrivener and Scrivener Publishing for their work in bringing this book to publication.

Dr. Sandip Kunar
June 2025

1

Introduction to Functional Composite Materials

Sandip Kunar[1]*, Gurudas Mandal[2], Jagadeesha T.[3] and Pranav Charkha[4]

[1]*Department of Mechanical Engineering, Aditya University, Surampalem, Andhra Pradesh, India*
[2]*Department of Metallurgical Engineering, Kazi Nazrul University, Asansol, India*
[3]*Department of Mechanical Engineering, National Institute of Technology, Calicut, India*
[4]*School of Engineering & Technology, Pimpri Chinchwad University, Pune, India*

Abstract

Composite materials are vital for everything, from supporting modern nations to facilitating new inventions. The main benefits are their strength, durability, light weight, flexibility in design, and resistance to corrosion. These materials are utilized in modern industries including building, medicine, oil and gas, sports, transportation, and aerospace. This chapter covers an overview of composite materials, including the properties of polymer matrix, reinforcements, a basic approach to material selection, composite production procedures, composite properties, and applications.

***Keywords*:** Natural fiber, biopolymer, thermoset, thermoplastic, fiber, matrix, composite

1.1 Introduction

Nature contains composites. The long cellulose fibers that make up wood are joined by lignin to form a composite substance. It is formed by joining two or more materials that do not dissolve concurrently and have quite discrete qualities. The composite's distinctive qualities are a result

**Corresponding author*: sandip.sandip.kunar@gmail.com

Sandip Kunar, Pranav Charkha, Santosh Jaju and Harish Tiwari (eds.) Functional Composites: Role in Modern Engineering, (1–20) © 2025 Scrivener Publishing LLC

of the interaction of its various constituent components. These materials have been applied by people in various contexts from a long time ago. To construct sturdy and long-lasting structures, early Mesopotamian and Egyptians settlers exploited the blending of straw and mud at around 1500 BC, which is when these materials were first applied. A block of brick has a strong confrontation to bending, ripping, and squeezing because of the mixture of straw and mud. The ancient composite materials such as ceramics and boats were reinforced by straw [1]. In 1200 AD, the first composite bow was produced by combining wood, bone, and animal glue during the Mongols period. Birch bark was used to wrap and press the bows. These bows were accurate and strong. The supremacy of Genghis Khan's military was ensured in part by composite Mongolian bows. Many of the biggest developments were brought about by wartime requirements because of their advantages, which include strength and light load. Various materials were created during World War II and transitioned from research and development to real manufacturing [2].

The fiber-reinforced polymers (FRP) are a direct result of the advancement and demand for composite materials. Over 7.5 million pounds of glass fibers had been utilized in 1945 for a variety of goods, mostly for military uses. The composite materials gained popularity and expanded quickly into the 1950s. The pioneers of composites made a bold attempt to expand the use of composites into other industries, including transportation, construction, and aircraft. The public sectors acquired the idea about the benefits of FRP composites, particularly no corrosion of composite materials. In 1946, the first composite boat hull was issued for commercial use. In 1947, an entire car body was made of composite and put through testing. As a result, Chevrolet Corvette was created in 1953. Several innovative molding techniques, including compression molding and sheet molding, emerged with the onset of the automotive age. The two methods became the most widely used molding procedures in the automotive and other sectors. Manufacturing techniques like vacuum bag molding, pultrusion, and large-scale filament winding were created in the early 1950s. The greatest market for composite materials in the 1960s was the marine industry. The first carbon fiber was patented in 1961 and frequently accessible after a few years. The industry commenced to manufacture the composites in the 1970s. During this time, numerous new resins and reinforcing fibers were created for use in composite applications. The automobile industry overtook the maritime industry as the largest market in the 1970s, and it still retains that position presently. Composites were initially employed in infrastructure applications in Asia and Europe in the late 1970s and early 1980s. In the 1990s, Aberfeldy, Scotland, saw the installation of the first pedestrian bridge made entirely of composite materials. During this time, the first all-composite vehicle bridge deck was built in Russell and Kansas, while the first FRP-reinforced concrete bridge deck was built in McKinleyville, West Virginia. Applications for composites are still being found today. New composites are created by combining nanomaterials with enhanced fibers and resins. In the early 2000s, nanotechnology started to appear in business goods. To enhance the electrical and mechanical characteristics of polymers, bulk carbon nanotube is utilized as composite reinforcement [3].

This material sector is advancing continuously with a large portion. For instance, engineers can modify the design of composite components based on the performance requirements, where more strength is required in the component. Wind turbine blades are always modifying the design of blade size and demand smart composite materials. By choosing the proper matrix material, the engineers can also decide the attributes like corrosion resistance, resistance to chemicals, etc. The use of natural fibers as reinforcements in composites to surrogate the synthetic fibers has acquired popularity recently owing to more environmental awareness and the requirement for sustainable development [4–7]. An overview of composite material, characteristics of polymer matrix, reinforcements, the fundamental method for choosing materials, techniques for producing composites, composite properties, and application are covered in this chapter.

1.2 Overview

As the most used word, composite materials are made up of two or more components, each of which exhibits a wide range of physical and/or chemical properties. A new material having features distinct from the individual component is fabricated by combining two or more fundamental materials. Blending two or more elementary materials produces a new substance with different properties from the constituent parts. Because the principal components remain discrete and the formation of structure is separate, composites must be appropriately separated from the solutions of solids and material mixtures [8].

Individual basic components, sometimes known as constituent materials, make up composite materials. The matrix, often known as the "binder," and the reinforcement are the two elementary types of basic materials that are identified. To make a composite, at least one substance from each category is needed. By maintaining the relative positions of the reinforcements, the matrix phase surrounds, envelops, and supports them. By adding their unique mechanical and physical resources, the reinforcements improve the matrix's qualities. While the infinite variety of reinforcements and binders allows the designer to create the best possible combinations, resulting in custom-made composites, the synergism between the two phases produces the quality materials [9].

The following are familiar instances of composite materials:

- Wood (cellulose fibers encased in lignin and hemicellulose)
- Bones (apatite, a hard mineral, collagen, and a soft protein)
- Pearlite is a mixture of cementite and ferrite [10, 11]

The following composite materials are categorized as follows:

- The matrix (binder) component serves as the basis for the first classification criterion. Ceramic matrix composites (CMCs), organic matrix composites (OMCs), and metal matrix composites (MMCs) are the three primary composite groups. In general, OMC refers to polymer matrix composites (PMCs) and carbon–carbon composites.
- Fiber-reinforced composites (FRCs), laminar composites, and particle composites are differentiated based on the second categorization criterion, which relates to the reinforcing phases. FRC can be further divided into types with reinforcements that are continuous or discontinuous fibers, respectively.
- FRC is made up of fibers encased in matrix materials. If the composite's characteristics rely on the fiber length, it is referred to as a discontinuous fiber composite or a short fiber composite. However, the composite is contemplated "continuous fiber reinforced" when the fiber length expansion does not cause the elastic modulus of the composite to grow further. Despite often having good tensile qualities, fibers are typically tiny in diameter and readily twist when compressed axially. The fibers must be reinforced for resisting the buckling.
- The particulate composites are made of particles, which can be powdered or flakes, that are dispersed in a binding matrix. Particle boards are fabricated of wood and concrete is good example of this form [12].

There are numerous further categories of composite materials, including the following:

(a) Grouping based on the kind of matrix materials:

➢ Composites made of metal (MMCs)

Despite having a relatively high specific mass, metal fibers are typically inexpensive. They are used for metal matrix reinforcement. This material has not more requirement because of their high density. The excellent fiber-matrix compatibility allows for the primary function in the metal-metal composite preparation. Metal matrices are reinforced with carbon steel fibers to endure temperatures up to 300°C. Fibers are composed of heat-resistant metals such as tungsten or molybdenum, are used to strengthen metal matrices. So they can tolerate higher temperature. The following fibers are used generally.

- Steel: regularly contains alloys of strengthening aluminium.
- Tungsten: used to reinforce materials that can withstand heat.
- Boric: extremely light but solid and inflexible; manufacturing is not simple. Boric fibers should be mentioned as a typical example, where a thin SiC layer is first applied to the surface of a thin tungsten wire to protect it from oxidation and boron diffusion into the matrix, followed by a boron layer that is chemically deposited on the wire by BCl3 vapor.

➢ Composite materials with an inorganic non-metallic matrix.

➢ Polymer matrix composites (PMCs)

Because of their low compactness, easy manufacturing, and desired technical properties, these polymers exhibit appropriate matrix materials. Consequently, polymeric resins endure elevated temperature. Due to that it is commonly applied in aviation [13]. Two important categories of polymers are thermosets and thermoplastics. A well-bonded 3D molecular structure that forms after curing, is what distinguishes thermosets. At high temperatures, some substances break down rather than melt. To amend this circumstances for curing and determining other characteristics, the key composition of the resin only needs to be replaced. They are also retained for long periods of time in a partially healed state. Additionally, thermosets have a great extent of flexibility. Therefore, they make excellent matrix bases for FRC in sophisticated applications. To create chopped fiber composites, thermosets are frequently utilized, particularly when utilizing a molded solution with good quality fibers. Thermoplastics are molecules with 1D or 2D structure that dissolved at greater temperatures and usually have inflated melting points. Another benefit is that their softening at high temperatures is reversible, meaning that cooling can return them to their previous characteristics. This makes it easier to apply well-established compression procedures for creating molded compounds. Thermoplastic-reinforced resins are currently a constant growing class of composites. These days, a lot of research efforts in this field are focused on improving the resins' main qualities and maximizing their performing benefits for certain applications. This includes attempts to use die-casting techniques to replace problematic metals. The reinforcement significantly alters the morphology of crystalline thermoplastics, causing the reinforcement to be stimulated to permit nucleation.

These resins can alter their creep characteristics across a wide temperature range, regardless of whether they are crystalline or amorphous. Reinforcement in such systems can improve the creep resistance and failure load, but this temperature range also covers the point at which resin usage is compromised [14, 15].

(b) Categorization based on the type of scattered phase:
- Continuous composites reinforced with fibers
- Textile fibers and woven-reinforced composites
- Composites reinforced with sheets
- Composites reinforced with very short fibers (sometimes known as "whiskers")
- Composites reinforced with particles
- Composites enhanced with nanoparticles

(c) Categorization based on reinforcing fiber type:

Composite materials made of carbon and graphite fibers are common traits of various materials:
- Less elongation at break than glass fibers.
- Tensile strength is lower than glass fibers at normal temperature.
- It is steady, chemically inert and excellent thermal performance up to 1000°C, and stable up to 2000°C when oxidation protected.
- Minimal expansion of heat, occasionally even contraction of heat.
- Significantly more resilient to fatigue than glass.
- Conductive of electricity.
- One hundred times the price of glass.
- Anisotropy is high.

1.3 Characteristics of Composites

The fact that there are numerous varieties of composite materials is already known to us based on their classification. Generally, the performance of various composites varies. However, they also share certain traits. When compared to well-familiar materials such as metals, polymer matrix composites have the subsequent unique properties:

(i) High specific strength and high specific modulus
High specific strength and specific modulus are the two main advantages of polymer matrix composites. The ratio of strength to density is known as specific strength, and the ratio of modulus to density is known as specific modulus. Length is the suitable dimension for both cases. These attributes are essential for structural materials used in aircraft. Because glass fibers have a relatively low modulus and a high density, glass fiber resin matrix composites have a somewhat lower specific modulus than metallic materials.

(ii) High damage resistance and rapid fatigue resistance
Often, there is no obvious sign of the severity of damage before metallic materials fatigue. In composites, the fiber/matrix interaction can prevent the spread of cracks. Those fiber links that are prone to breaking are always where fatigue failure begins. There is a significant forerun before the start of the last destruction because crack growth or destruction spreads gradually over an extended period. The S-N curve of fatigue properties shows that the fatigue strength of most metallic materials is only 40–50% of tensile strength, whereas for carbon fiber/polyester composites, this value rises to 72–82% for glass fiber composites. The percentage falls between these two ranges.

(iii) Good damping properties
The inherent frequency of composites is high, and it is generally difficult to generate a resonance. Simultaneously, composites have a strong vibration damping because the fiber/matrix contact readily absorbs vibrational energy. It is simple to halt vibrations if they do occur [16].

(iv) Practical processing methods
Fiber matrix can be preferred based on the product's performance needs and usage circumstances. As a result, custom materials can be created upon request. Depending on the product's size, shape,

and quantity, different molding processing procedures might be used. By reducing the number of separate parts, integrated molding can save weight, time, and material.

1.4 A Fundamental Method for Choosing Materials

When creating a novel product, choosing the right material for a planned use is crucial and extremely important. The performance of the finished product will fulfil the specified function and performance requirements based on the choice of the best material. Materials that are inappropriate or less suited can lead to the following problems:

- Production-related technological issues.
- Increasing production costs, which raises the final product's price.
- Adverse ecological effects.
- The choice of materials.

The process of choosing materials is intricate and varied, requiring consideration of several criteria, including:

- Material costs.
- Production costs.
- Energy and raw material demand (process material intensity).
- The potential effects of material choice on the environment, which are influenced by the life cycle of production and consumption.
- Material recycling accessibility.

The product's structure (its shape), material, technology, and functioning (based on its individual components) are all strongly related to one another and cannot be considered separately; as a result, the technology cannot be chosen separately from the material.

- o Requirements and parameters for product development and design:
    - The selection of the best specific elements based on their performance in the future.

- Evaluation of component compatibility: every stage of the composite material must maintain its advantageous properties. The constituent parts cannot harm or negatively impact one another.
- Selecting a geometrical form that is appropriate for every stage.
- The distribution of composite phases should allow for their synergistic operation.

1.5 Polymer Matrix

It comprises three parts: (i) matrix; (ii) reinforcements, and (iii) fine interphase region [17, 18]. The engineers can modulate the attributes to satisfy particular needs by carefully choosing the matrix, reinforcement, and manufacturing system that combines them [19]. Numerous new composites, some with extremely desirable features, have been produced in recent decades. For composites, any material can be used as the matrix. Nonetheless, metals, polymers, and ceramics are typically used as matrix materials. Polymers are commonly available in the composites market. A variety of polymer matrices are available for use in composite materials. Thermoset composites are more universal than thermoplastic composites among polymer matrix composites. Even though their same names, thermoset and thermoplastics have completely distinct qualities and applications. The composite product designs can be achieved by being aware of the working variations [20, 21].

There is a market for both thermoset and thermoplastic materials. Broadly speaking, thermosets often have a lengthy history, a solid market position, lower raw material costs, and the ability to easily wet reinforcing fiber and shape final constituent structures. Thermosets are commonly easier to work with than thermoplastic. Compared to thermoset, thermoplastics are typically more resilient or less brittle. They can be more simply reused, restored, and have greater chemical resistance.

Polyester resins, polyurethane, and epoxy resins are the several types of thermosets. One of the main parameters compelling the thermoset market is the quick industrialization of developing nations worldwide. The global thermoset market is projected to grow owing to the need for lightweight, high-performance materials from a variety of end-use sectors, including water tanks, chemical tanks, and the automotive sector. The market is projected to be driven by the expanding demand of thermosets. The GDP growth rates of the four BRIC countries are higher than the global GDP growth rate, making them the quickest-growing economy in the world.

Nonetheless, one of the main impeding market expansion is the regular volatility of raw material prices. Asia-Pacific is the largest market for thermosets because of the expansion of the automotive industry, particularly in India and China. Japan is a developed market that is anticipated to stay unchanged in the next years. China is the world's largest vehicle market, and India is among the top five global automobile markets. Asia is not only the largest thermoset market, but it is also the one with the quickest rate of growth. The USA's regulatory goal to reduce car weight by 50% by 2020 to minimize fuel consumption is the main factor driving the thermosets market in North America. The two most broadly used thermoset types on the international market are polyester resins and polyurethane. Large multinational firms that operate throughout the value chain control most of the global thermoset market. Bayer AG, Arkema, and Eastman Chemical Company are a few of the significant businesses involved in the thermosets industry [22]. The industry has been the primary user of thermosets up to this point. For several reasons, thermosets are typically preferred, particularly on commercial aircraft. For thirty to forty years, thermoset composites have been employed in the aerospace industry. For instance, the Boeing 787's fuselage is made of a polymer based on epoxy [23].

However, the use of thermoplastic polymers, which are more readily moldable and resettable than thermoset polymers, is a material trend that is expanding in the fiber-reinforced polymer industry. The American Composites Manufacturers Association projects that the thermoplastics sector would expand by 5% over the coming years, reaching a projected $8 billion by 2017. Emerging markets are predicted to offer even greater growth potential. When it comes to the use of composite materials, thermoplastic polymers also provide a simple solution for recycling composite components. Thermoplastic composites have an essentially endless shelf life and add value for sectors that care about composite recyclability since they can repeatedly repeat the heating and cooling cycle. This is particularly true for the expansion of thermoplastics made of natural fibers in United States and Western Europe. Because the production of thermoplastic automotive components is expanding rapidly in Eastern Europe and Asia, those nations will lead the growth in thermoplastic adoption, according to Lucintel, a leading worldwide management consulting and market research business. Though they might not grow as quickly, the automotive industries are predicted to increase gradually over the next five years, primarily because of the adoption of innovative composite applications.

According to the report, there is a lot of potential in developing nations like China, Russia, Brazil, and India, even if profits will be constrained by growing energy prices and competition from cheaper materials [24]. The usage of thermoplastics rather than "traditional" thermoset epoxies has increased recently, which is a significant development in the aerospace sector. Additionally, this presents a chance for thermoplastics.

In general, thermoplastics are the most common type of plastic, particularly for nonreinforced applications. Thermosets have a specific use in non-reinforced applications where their special feature gives them an advantage. Thermoset and thermoplastic, on the other hand, are only utilized in areas where their distinct advantages are significant in the reinforced or composites market. About 80% of the materials utilized in the composites sector are thermoset [25]. The value of the global composite resin market by end-use applications was USD 9317.4 million in 2014, and it is estimated to increase at a compound annual growth rate of 5.6% from 2015 to 2020 [26].

1.6 Reinforcements

Composite reinforcements have a variety of sizes and shapes, including particles, flakes, and fibers. Fibers have the greatest impact on the characteristics of the composite materials and are the form most frequently utilized in composite applications. These include the capacity to process and construct the composite part in a variety of shapes using different processes. Glass fibers are utilized as reinforcement, most notably by Venetian glassmakers of the Renaissance. An Owens-Illinois and Corning Glass joint venture began producing commercially significant continuous glass fiber filaments in 1937. As shown below, a range of glass fiber compositions are accessible for various applications.

- Originally constructed from window glass, grade A is high-alkali-grade glass.
- Grade C is glass that is resistant to corrosion and acidic conditions.
- Grade D is low-dielectric glass, which is made of quartz and has good radar transparency.
- The most popular degree of reinforcement is electrical insulation, or grade E.
- Glass with a high modulus is grade M.

- S2-glass is a popular variation of grade S, which is high-strength-grade glass. Compared to E-glass, this fiber has a greater Young's modulus and is more temperature resistant. Additionally, it costs a lot more.

The degree of alignment perfection determines the modulus of carbon fibers. Complex voids that are stretched parallel to the fiber axis due to alignment errors serve as stress raisers and weak spots. The manufacturing route and conditions have a significant impact on the alignment. High-strength carbon fibers are made up of tiny crystallites of "turbostratic" graphite and have diameters of 7 to 8 μm. The graphite's layer planes must line up parallel to the fiber axis to achieve high modulus and strength [27]. High stiffness, temperature, and chemical resistance are the advantages of carbon fibers. Applications for carbon fibers are numerous and include consumer items, recreational goods, automotive, and aerospace.

The aramid fiber is recognized as Kevlar, which is lightweight and robust than steel and is exploited in bulletproof vests and other body armor worldwide, and was created in 1965 by DuPont chemist Kwolek. The amide (CONH) group and aromatic (aryl) benzene rings are changing in the materials' chemical structure. The fibers used for reinforcement are marketed under the names Twaron from AkzoNobel and Kevlar from DuPont.

Due to their favourable mechanical properties, low density, environmental benefits, renewability, and economic feasibility, natural fibers have recently drawn increased interest in composite applications. Natural fibers, which are categorized based on whether they originate from plant, animal, or mineral sources, are simply defined as those that are neither synthetic nor man-made [28]. Natural plant fibers only originate from vegetative sources and are fully biodegradable. Fiber-reinforced polymer matrices have garnered significant interest in a range of applications due to their advantageous properties. Current research suggests that industry interest in natural fiber composites will continue to grow quickly on a global basis.

1.7 Techniques for Producing Composites

The composite materials are produced in different ways. The materials, performance, part design, and final usage or application will all impact the process used for a given part.

1.7.1 Molding Using Open Contact

One open contact molding method for creating composite materials is hand lay-up. Resins are manually infused into fibers to create bonded, knitted, woven, or stitched textiles. This approach contains first treating the mold with mold release, then placing dry textiles or fibers over it, and finally pouring and spreading liquid resin over the fiber beds [29]. Naturally, rollers are utilized for this, but nip-roller-type impregnators which use rotating rollers, and a resin bath are increasingly being used to force resin into textiles. The fibers are moistened and air trapped in the lay-ups is expelled using a roller or brush. Following the wetting of a few fiber layers, laminates are allowed to cure in ambient air. Additional coats are applied once these layers have dried. Another open-mold composite application method is spray-up. The hand lay-up approach is thought to be expanded upon by the spray lay-up method. The mold is first treated with mold release in this process. After the mold release is used, a gel coat, if one is employed, is applied into the mold at a specific rate. After the gel coat is dried, the mold is prepared for processing. A chopper spray gun is utilized to spray on the fiber and catalyzed resin into the mold at a viscosity of 500–1,000 cps. To apply both materials to the mold's surface at the same time, the gun cuts continuous fiber into small bundle lengths and then blasts the short fibers straight into the stream of sprayed resin. The workers use rollers to manually condense the laminate in the last stages of the spray-up procedure. After curing and cooling, the composite portion is taken out of the mold [30, 31]. Hand lay-up and spray-up procedures are sometimes applied to lower the labor expenditures. It is commonly used to make bathtubs, truck cabs, and other rather large and simple designs.

1.7.2 Resin Infusion Method

The most often used acronym when discussing low pressure closed mold reinforced composite molding is probably vacuum assisted resin transfer molding (VARTM). When the composite is molded using a rigid mold to provide part geometry and a thin, flexible membrane over the fiber, the fiber is compressed tightly against the rigid mold surface by the outer atmospheric pressure. It is important to remember that VARTM only refers to applying vacuum to the molding tool's exit vent; therefore, any type of resin transfer molding that applies vacuum to the vent would be eligible for the VARTM association. The aerospace industry frequently uses rigid matched mold sets to mold very high fiber volume composites. The molds are clamped closed using a press, and resin injection pressures are usually less than 100 psi, but they can reach 600 psi when used to fill the mold and encapsulate the dry fiber inside [32–34]. Small fabrication companies such as Boeing Company and NASA have all used this technique to create aerospace-quality laminates without the need for an autoclave [35].

1.7.3 Injection Molding

This is most frequently applied to filled thermoplastics, like nylon with chopped glass fiber, and it is quick, high volume, and low pressure. For about 150 years, this method had been used. Since the introduction in the 1960s, reciprocating screw injection molding machines have remained in use [36]. Some multiple cavity molds may manufacture many components per hour, and the requirement of injection speeds is less.

1.7.4 Filament Winding

A continuous production technique with comparatively inexpensive material costs, filament winding may be greatly automated and replicated. The cylindrical mandrel is suspended horizontally between end supports. Dry fibers are wetted by passing them through a resin bath. Using the traverse carriage, the fiber application tool goes back and forth down the length of a revolving mandrel, applying fiber to the tool in a preset pattern. The axes of motion are arranged using computer-controlled filament-winding machines [37–39]. One type of aeronautical composite materials is filament winding.

1.7.5 Pultrusion Process

Fiber-reinforced polymer structures can be produced in continuous lengths using the composite pultrusion processing method. A continuous bundle of dry fiber is drawn through a hot resin-wetting station throughout this procedure. After the wet bundle is drawn into heated dies, the dies create the pulled fiber's cross-sectional shape. The composites are created when the resin has dried. The long-cured object is subsequently cut into parts. This method is only applicable to straight components with a fixed cross-section, including ladder rails, frame sections, T-beams, and I-beams. Ladders and moldings are examples of linear components made *via* pultrusion.

1.7.6 Additive Manufacturing

This method is used to create a physical product from 3D digital model, commonly using many thin layers of a material one after the other. Polymer composite filament is used to create composite structures with a single nozzle. This more contemporary method of producing composite parts emerged from attempts to lower costs throughout the design and product development, with a focus on the labor, material, and time-intensive tool-making process [40–43].

1.8 Composite Properties

Generalization is challenging since there are a growing several materials that fit for different composites. The composites have the pertinent characteristics such as resistance to wear, high modulus, corrosion resistance, and high strength. Carbon matrix composites typically exhibit exceptional resistance to wear, oxidation, and corrosion in addition to their capacity to resist high temperatures. The strengths of monolithic ceramics exhibit significant scatter because they are fragile materials with moderately low fracture toughness. For applications that are exposed to high levels of torsional stress, they are not practical materials. Ceramics can become attractive materials for higher temperature applications such as aviation engines, by being reinforced with continuous fibers. The lightweight nature of the polymer composite materials improves the structural stability and fuel efficiency of cars made of composites. They also provide enhanced heat resistance and a high strength-to-weight ratio. Composites have greater qualities and uses based on the kind of matrix, reinforcement, processing, etc. One of the key elements in achieving better fiber reinforcement polymer composite is thought to be the strength of the link between polymer matrix and fiber. Increased applications in aerospace, defense, and transportation are responsible for the market expansion for composites. The world-wide market for composite materials is projected to develop at a compound annual growth rate from 2019 to 2024, reaching an estimated $42 billion. By 2024, the international market is expected to raise to a value of $120 billion [44].

These materials are widely used in constructions due in large part to their exceptional mechanical qualities. Composites are chosen for a growing range of applications, particularly electronics and photonics heat management, because to their significant physical characteristics, particularly their high thermal conductivity. The extraordinarily high heat conductivity of certain composites is another distinctive and crucial characteristic. Consequently, these are being applied more often in situations where heat dissipation is a vital design factor. Moreover, the low density of composites makes them exclusively beneficial for thermal control applications where weight is crucial, such as avionics, tablets, and phones.

1.9 Latest Developments

Advanced manufacturing techniques have been made possible by the development of contemporary composites.

- Carbon fiber (1961): Another significant advancement in composite technology was made in 1961 with the patenting of carbon fiber. Because of its rigidity and high strength-to-weight ratio, carbon fiber was immediately used in the sporting goods, automobile, and aerospace industries.
- Kevlar (1966): The development of Kevlar, an aramid fiber with a high tensile-strength-to-weight ratio, by DuPont chemist Stephanie Kwolek created new opportunities for high-performance composites and ballistic protection.
- Early in the new millennium, nanotechnology was used into composite materials, which resulted in notable gains in strength, durability, and multipurpose. When compared to conventional composites, nanocomposites integrate nanoparticles into the matrix, which can show noticeably better qualities.
- Composite 3D printing (2014): In 2014, Mark Forged unveiled the first carbon fiber 3D printer in history, marking a breakthrough in composite additive manufacturing. With the use of this technology, intricate, highly durable components with ideal fiber orientation can be produced.
- Innovation in advanced composites manufacturing (2015): A dedication to improving composite technology and manufacturing techniques was demonstrated with the establishment of the Institute for Advanced Composites Manufacturing Innovation through the amount of $259 million public–private partnership. The goal of this project is to hasten the creation and uptake of innovative composite materials and procedures [45].

1.10 Applications

- Spacecrafts including solar reflectors, satellite structures, rocket engines, radar, antenna structures, and more
- Aircraft components including jet engines, helicopter transmission structures, wing box structures, turbine blades, and shafts
- Automotive components include abrasive materials, electrical machinery, truss members, pressure vessels, cutting tools, electrical brushes, and engine parts such as cylinders, pistons, and bearing materials
- Specifically, carbon–wood epoxy composite is utilized for the production of wind turbine blades.

1.11 Conclusion

One of the numerous benefits of composites is that they may be made from a variety of material combinations, giving designers more creative freedom. Additionally, the composites are easily moldable into intricate shapes. The materials are fabricated to meet the requirements. Composites have less weight than most metals and woods have a lower density than many metals. Compared to certain other materials, they are more robust. The materials are resistant to weather-related and strong chemical damage. Composites are low maintenance and have a long service life. The design will be varied for different composite products due to various matrix and their forms, dissimilar reinforcements, and manufacturing techniques. Consequently, the selection of production method and composite material depends on the requirements of products in rural societies. Research on composite materials is still ongoing. Several factors are to be considered to increase the advantages of the composites such as material formulation and selection, material design, concept creation, product manufacturing, and market.

References

1. Johnson, T., History of composites. The evolution of lightweight composite materials, 2018, Available from: https://www.thoughtco.com/history-ofcomposites-820404.

2. Available from: https://www.marbal.com/language/en/applications/history-of-composites/
3. Available from: http://compositeslab.com/composites-101/history-ofcomposites/
4. Yuanjian, T. and Isaac, D.H., Impact and fatigue behaviour of hemp fibre composites. *Compos. Sci. Technol.*, 67, 3300–3307, 2007, DOI: 10.1016/j.compscitech.2007.03.039.
5. Sanjay, M.R., Arpitha, G.R., Naik, L.L., Gopalakrishna, K., Yogesha, B., Applications of natural fibers and its composites: An overview. *Nat. Resour.*, 7, 108–114, 2016, DOI: 10.4236/nr.2016.730.
6. Ho, M.P., Wang, H., Lee, J.H., Ho, C.K., Lau, K.T., Leng, J.S., *et al.*, Critical factors on manufacturing processes of natural fibre composites. *Composites, Part B*, 43, 3549–3562, 2012, DOI: 10.1016/j.compositesb.2011.10.001.
7. Sathishkumar, T.P., Navaneethakrishnan, P., Shankar, S., Tensile and flexural properties of snake grass natural fiber reinforced isophthallic polyester composites. *Compos. Sci. Technol.*, 72, 1183–1190, 2012, DOI: 10.1016/j.compscitech.2012.04.001.
8. Hashin, Z., Analysis of composite materials—A survey. *J. Appl. Mech.*, 50, 3, 481–505, 1983.
9. Hon, D. and Shiraishi, N. (Eds.), *Wood and Cellulose Chemistry*, 2nd ed., p. 5, Marcel Dekker, New York, 2001.
10. Ibrahim, I.D., Jamiru, T., Sadiku, R.E., Kupolati, W.K., Agwuncha, S.C., Ekundayo, G., The use of polypropylene in bamboo fibre composite and their mechanical properties—A review. *J. Reinf. Plast. Compos.*, 34, 16, 1347–1356, 2015.
11. Altenbach, H., *et al.*, *Mechanics of Composite Structural Elements*, Springer-Verlag, Berlin, 2004.
12. Miracle, D.B. and Donaldson, S.L., Introduction to composites, in: *ASM Handbook*, vol. 21, pp. 3–17, 2001.
13. Wang, R.-M., Zheng, S.-R., Zheng, Y.-P., *Polymer Matrix Composites and Technology*, WP, Woodhead Publishing, Oxford, 2011.
14. Thostenson, E.T., Ren, Z., Chou, T.-W., Advances in the science and technology of carbon nanotubes and their composites: A review. *Compos. Sci. Technol.*, 61, 13, 1899–1912, 2001.
15. Das, B., *et al.*, Nano-indentation studies on polymer matrix composites reinforced by few layer graphene. *Nanotechnology*, 20, 12, 125705, 2009.
16. Christensen, R.M., *Mechanics of Composite Materials*, Mineola, New York, 2005, 2012.
17. Kalpakjian, S. and Schmid, S.R., *Manufacturing Engineering and Technology*, International Edition. 4th ed., Prentice Hall, Inc., 2001, ISBN 0-13-017440-8.
18. Cantwell, W.J. and Morton, J., The impact resistance of composite materials—A review. *Composites*, 22, 5, 347–362, 1991, DOI: 10.1016/0010-4361(91)90549-V.

19. Williams, J., The Science and Technology of Composite Materials, Australian Academy of Science, 2019, Available from: https://www.science.org. au/curious/technology-future/composite-materials.
20. Available from: http://www.modorplastics.com/thermoset-vs-thermoplastics.
21. Available from: http://www.automateddynamics.com/article/compositebasics/thermoset-vs-thermoplasticcomposites.
22. Available from: http://www.transparencymarketresearch.com/thermosetsmarket.html
23. Available from: http://articles.sae.org/8547/
24. Available from: http://compositesmanufacturingmagazine.com/2012/07/exponential-growth-expectedthermoplastic-composites/
25. Brent Strong, A., Fundamentals of composites manufacturing, in: *Materials, Methods, and Applications*, 2nd ed., Society of Manufacturing Engineers, USA, 2008.
26. Available from: http://www.marke tsandmarkets.com/Market-Reports/composites-resin-market-66717271.html
27. Available from: https://www.fose1.plymouth.ac.uk
28. Asokan, P., Firdoous, M., Sonal, W., Properties and potential of bio fibres, bio binders, and bio composites. *Rev. Adv. Mater. Sci.*, 30, 254–261, 2012, ISSN: 1605-8127.
29. Available from: https://netcomposite s.com/guide/manufacturing/wet-handlay-up/
30. Available from: https://www.compositesworld.com
31. Available from: https://www.eppcomposites.com/spray-layup-process.html
32. Heider, D., Graf, A., Fink Bruce, K., Gillespie, J.W., Jr, Feedback control of the vacuum-assisted resin transfer molding (VARTM) process, in: *Process Control and Sensors for Manufacturing*, vol. II (3589), pp. 133–141, 1999, DOI: 10.1117/12.339956.
33. Hayward, J.S. and Harris, B., The effect of vacuum assistance in resin transfer moulding. *Compos. Manuf.*, 1, 3, 161–166, 1990, DOI: 10.1016/0956-7143(90)90163-Q.
34. Lih, T.C., Azm, A.I., Muhammad, N., Delamination and surface roughness analyses in drilling hybrid carbon/glass composite. *Mater. Manuf. Processes*, 31, 1366–1376, 2016, DOI: 10.1080/10426914.2015.1103864.
35. Qi, B., Raju, J., Kruckenberg, T., Stanning, R., A resin film infusion process for manufacture of advanced composite structures. *Compos. Struct.*, 47, 1–4, 471–476, 1999.
36. Frizelle, W.G., Injection molding technology, in: *Applied Plastics Engineering Handbook*, M. Kutz and W. Andrew (Eds.), Elsevier, USA, 2011.
37. Minsch, N., Herrmann, F.H., Gereke, T., Nocke, A., Cherif, C., Analysis of filament winding processes and potential equipment technologies. *Procedia CIRP*, 66, 125–130, 2017, DOI: 10.1016/j.procir.2017.03.284.

38. Ma, Q., Rejab, R., Kaige, J., Idric, M., Harith, M.N., Filament windingtechnique, experiment and simulation analysis on tubular structure. *IOP Conf. Ser.: Mater. Sci. Eng.*, 342, 012029, iCITES2018, 2018, DOI: 10.1088/1757-899X/342/1/012029.
39. Tarnopol'skii, Y.M., Peters, S.T., Beil, A.I., Filament winding, in: *Handbook of Composites*, S.T. Peters (Ed.), Springer, Boston, MA, 1998, DOI: 10.1007/978-1-4615-6389-1-22.
40. Bryll, K., Piesowicz, E., Szymański, P., Ślączka, W., Pijanowski, M., Polymer composite manufacturing by FDM 3D printing technology. *MATEC Web Conf.*, 237, 02006, 2018, DOI: 10.1051/matecconf/2018237020.
41. Dudek, P., FDM 3D printing technology in manufacturing composite elements. *Arch. Metall. Mater.*, 58, 4, 1415–1418, 2013.
42. Apparatus and method for creating three-dimensional objects. Patent US 5121329 A, 1992.
43. Ning, F., Cong, W., Qiu, J., Wei, J., Wang, S., Additive manufacturing of carbon fiber reinforced thermoplastic composites using fused deposition modeling. *Composites, Part B*, 80, 369–378, 2015.
44. Available from: www.plasticomp.com
45. Available from: https://www.addcomposites.com/post/introduction-to-composite-materials

2

Shape Memory Alloys as Functional Composites

Param Singh*, Yatin Khanna, Roopak Varshney and Yajush Walia

Department of Mechanical Engineering, National Institute of Technology Hamirpur, Hamirpur, India

Abstract

Shape memory alloys (SMAs) have become highly useful materials with distinctive features in a variety of engineering and scientific fields. This book chapter investigates the incorporation of SMAs into functional composites, opening up a wide range of potential novel uses. This chapter digs into the synthesis and processing methods for inserting SMAs into composite materials, starting with a review of SMAs and their extraordinary shape memory effect. Researchers have developed composites with improved functions, such as shape memory, mechanical strength, and thermal stability, by fusing the extraordinary qualities of SMAs with other materials like polymers, metals, or ceramics. This chapter also examines scalability, manufacturing methods, and performance optimization as well as the difficulties and potential of SMA-based composites. The article makes clear how SMAs have the capacity to transform sectors and open the door for cutting-edge technical developments. Finally, "Shape Memory Alloys as Functional Composites: Innovations and Applications" provides a thorough examination of the interaction between SMAs and composite materials, providing insightful information on their synthesis, uses, and future directions. This chapter will serve as a vital resource for scientists, engineers, and creators in their quest to utilize SMAs to the fullest in the creation of cutting-edge functional composites.

Keywords**:** Shape memory effect, super elasticity, austenite phase, martensite phase, aerospace, medical devices, automotive sector, shape memory alloys

**Corresponding author*: psingh@nith.ac.in

Sandip Kunar, Pranav Charkha, Santosh Jaju and Harish Tiwari (eds.) Functional Composites: Role in Modern Engineering, (21–48) © 2025 Scrivener Publishing LLC

2.1 Introduction

An overview is given of shape memory alloys, often known as SMAs, which are a special family of materials that have the extraordinary capacity to revert back to their original form after being significantly deformed as a result of being exposed to particular stimuli from the outside world. The underlying phase transitions that take place inside the microstructure of the material are what are responsible for this unusual behavior.

2.1.1 Exploration and Attempts at Development

Arne Olander, a Swedish metallurgist, made the discovery of the shape memory effect in the 1930s when he was working with gold–cadmium (Au–Cd) alloys and made an unintentional observation of the shape memory behavior. However, it was not until the 1960s when researchers at the Naval Ordnance Laboratory in the United States developed nickel–titanium (NiTi) alloys. This invention brought to a major increase the amount of attention paid to the phenomena.

Qualities and behavior: Shape memory alloys have a number of distinct properties that make them desirable for use in a diverse array of applications, including the following:

- ***Shape memory effect (SME):*** The ability of shape memory alloys to "remember" and revert back to their initial shape after being subjected to certain heat or stress-induced circumstances is the quality that most distinguishes SMAs from other types of materials. This effect is the result of a reversible solid-state phase transformation that takes place between two separate crystalline phases known as austenite and martensite. When a shape memory alloy (SMA) is in its high-temperature austenite phase, it is much more malleable and can be easily manipulated into a different shape. The material will convert into a martensitic phase when it cools, and it will maintain the shape it was in before it was distorted. Nevertheless, if the material is heated once more, it will transform back into its austenitic state, returning to its initial shape with a high degree of accuracy [1].
- ***Super elasticity:*** Super elasticity, also known as pseudo elasticity or rubber-like behavior, is yet another extraordinary trait displayed by SMAs. Super elasticity is often referred to

> as elastomeric activity. SMAs are able to withstand significant elastic deformation and restore their original shape after being unloaded, in contrast to traditional metals which are unable to retain their original form when subjected to stress that is greater than their yield point. This behavior is the result of a reversible martensitic transition that was triggered by stress. During super elastic deformation, the stress–strain curve of SMAs demonstrates a characteristic plateau zone. In this region, the material demonstrates a practically constant stress over a vast strain range. Due to their one-of-a-kind behavior, SMAs are particularly well-suited for use in applications that call for high flexibility, vibration dampening, and energy absorption [2].

As a result of their extraordinary qualities and wide range of potential applications, shape memory alloys have caused a revolution in the field of materials engineering. The shape memory effect and super elasticity displayed by SMAs present one-of-a-kind chances for the development of creative solutions in a variety of different disciplines. The potential for new discoveries and uses of shape memory alloys is quite large as research and development in this sector continues to make progress.

2.2 Composition and Microstructure of Shape Memory Alloys' (SMAs') Composition

The composition of SMAs include nickel (Ni), titanium (Ti), copper (Cu), and iron, which are the four metals that are utilized most frequently in the production of alloys. Typically, alloys are formed of one or more metallic elements (Fe).

These constituents, when mixed together in predetermined ratios, give rise to alloys with the characteristics that have been specified. The nickel–titanium (NiTi) alloy, more commonly referred to as nitinol, is the shape memory alloy that has been subjected to the most research and sees the most amount of use. Alloys made of nickel and titanium are known as nickel–titanium (NiTi) alloys. NiTi alloys are renowned for their extraordinary qualities of shape memory and super elasticity. The proportion of nickel (Ni) to titanium (Ti) in NiTi alloys can range anywhere from around 50 to 55 at%, with the percentage of nickel (Ni) being constant (Ti).

The alloy is able to undergo reversible phase changes thanks to this particular composition, which also enables it to display the shape memory and super elastic characteristics that are needed. Additional alloying components in addition to nickel and titanium: SMAs may also include additional alloying components in order to vary the properties of the material.

Copper (Cu), when added to NiTi alloys, can significantly improve the shape memory effect of the combination. Copper was added to the alloy, which caused the transformation temperatures to rise and led to an improvement in the alloy's thermal stability.

In addition to NiTi alloys, iron-based SMAs, such as Fe–Mn–Si alloys, have attracted a lot of attention recently due to their lower cost, better strength, and environmental friendliness in comparison to NiTi alloys. These alloys typically include the elements iron (Fe), manganese (Mn), and silicon (Si), along with trace amounts of a few other elements that are added to help maximize the properties of the alloy [3].

2.2.1 Microstructure

The microstructure of shape memory alloys (SMAs) is an important factor in the mechanical and shape memory capabilities of these materials.

SMAs typically display a dual-phase microstructure that is composed of the following phases:

- ***Austenite phase:*** The austenite phase is the high-temperature phase of the structural metal alloys (SMAs). The desired effect of shape memory is provided by a crystal structure known as face-centered cubic, abbreviated FCC for short. The extreme ductility and capacity to undergo reversible deformation that austenite possesses are two of its defining characteristics.
- ***Martensite phase:*** The martensite phase of SMAs is the phase that exists at low temperatures. Crystallographically speaking, it has either a body-centered cubic (BCC) or a mixture of body-centered cubic and hexagonal close-packed (HCP) arrangements. Martensite is what allows distorted SMA structures to keep their original shape after being stretched. In comparison to austenite, it has greater strength but lesser ductility than the latter.

The shape memory effect in SMAs is caused by a reversible phase transformation between the austenite and martensite phases. This transformation can take place in both directions. Heating causes the martensite phase to convert back into austenite, which ultimately results in the shape returning to how it was before it was heated. The martensitic transformation, on the other hand, is induced by cooling, which also causes the shape to be retained after deformation [4].

Composition, temperature, stress, and strain are only few of the elements that can have an effect on the phase transformation behavior. The phase transition characteristics and the shape memory capabilities of SMAs can also be influenced by the presence of alloying elements, microstructural flaws, and heat treatment. Control of the microstructure: The ability to control the microstructure of SMAs is essential for modifying the properties of these materials. Figure 2.1 illustrates the phase transformation process in SMAs, showing the reversible transition between austenite and martensite phases under thermal stimuli.

SMAs can have their phase transformation temperatures, mechanical strength, and shape memory behavior optimised by the use of a variety of different techniques. Some of these techniques include heat treatment, thermomechanical processing, and alloying element additions.

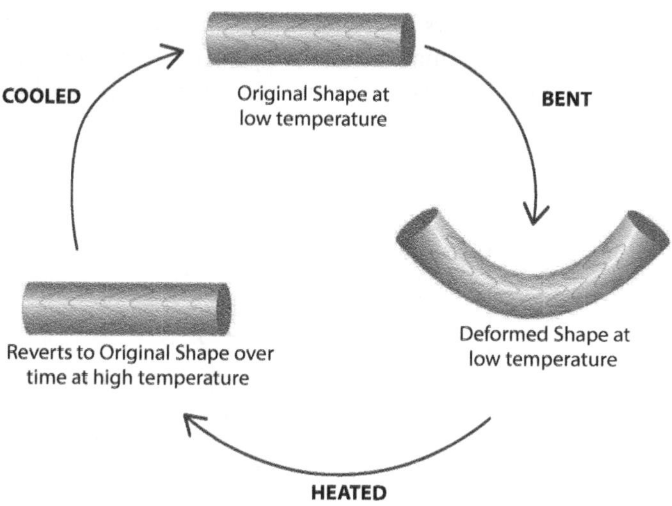

Figure 2.1 Phase transformation process for SMAs.

2.2.2 Characterization Techniques

Optical microscopy, scanning electron microscopy (SEM), transmission electron microscopy (TEM), X-ray diffraction (XRD), and differential scanning calorimetry are some of the techniques that can be used to describe the microstructure of SMAs (DSC). These methods offer extremely useful insights into the crystal structure, phase distribution, and microstructural characteristics of SMAs, all of which contribute to the shape memory behavior of these materials.

It is absolutely necessary to have an understanding of the composition as well as the microstructure of SMAs in order to customise their properties and create useful composite structures. Researchers are able to improve the shape memory effect, super elasticity, and other features of shape memory alloys (SMAs) that are desirable for certain applications by modifying the parameters listed above.

2.2.3 Shape Memory Effect

The shape memory effect, often known as the SME, is a remarkable and fascinating quality that can be seen in shape memory alloys (SMAs). It refers to the ability of certain alloys to "remember" and restore their previous shape after suffering severe deformation as a result of being subjected to particular external stimuli, such as temperature or stress. In other words, it is the ability of these alloys to "remember." The shape memory effect is caused by a phase transformation that may be reversed inside the SMA microstructure, which occurs between two separate crystalline phases

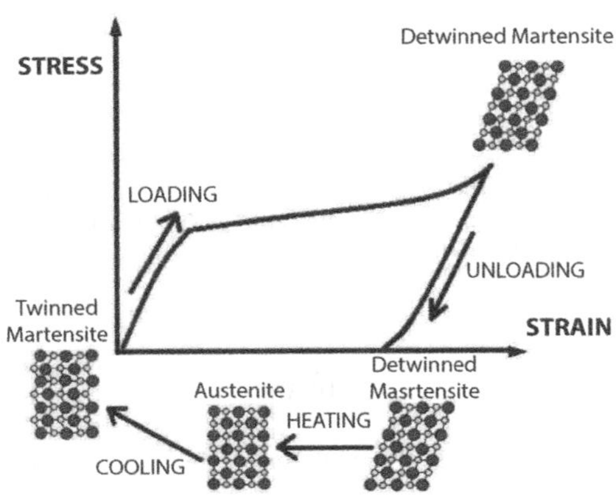

Figure 2.2 Shape memory effect.

known as austenite and martensite [5]. Figure 2.2 depicts the stress-strain behavior during the shape memory effect.

- ***Austenite phase:*** SMAs are known to exist in the austenite phase at higher temperatures. This phase has a crystal structure known as face-centered cubic (FCC), and it is defined by its cubic faces. During this stage, the material will have a higher ductility, meaning that it will be easier to deform and mold it into a variety of different forms.
- ***Martensite phase:*** When the temperature of the SMA falls below a certain threshold, referred to as the martensite start temperature (Ms), the material goes through a phase transformation and changes into the martensite phase. Crystal structure of martensite can be described as either body-centered cubic (BCC) or a blend of body-centered cubic and hexagonal close-packed (HCP).

  In most cases, the beginning of this change is brought on by the application of mechanical stress or a reduction in temperature. The martensite phase is what allows the distorted SMA structures to keep their original shape after being stretched. It is possible for the material to maintain the temporary shape even after the deformation has been eliminated because the material possesses a unique feature that is referred to as shape memory. The martensitic condition in which the distorted structure exists is referred to as the "parent" state.
- ***Reversible transformation:*** The shape memory effect is shown when the shape memory alloy (SMA) has been heated to a temperature that is higher than the austenite finish temperature (Af). In the course of this heating process, the martensite phase undergoes a transformation that returns it to its original austenite state.

  This transformation in the opposite direction causes the material to revert to the state in which it was "learned" or "remembered," which allows the original shape of the material to be restored. As the temperature rises, the atoms that make up the crystal lattice are forced to reorganise themselves, which results in the transformation going in the other direction.

  Even after undergoing significant deformations, the crystal structure of the austenite phase is brought back to its original state, and the SMA recovers its initial shape with a high degree of accuracy. The reverse transformation is generally driven by thermal energy, and it can be initiated by heating the SMA through a variety of methods, such as resistive heating, electrical current, or exposure to a suitable heat source. In general, thermal energy is what drives the transformation, and it can be initiated by heating the SMA.

- ***Temperature-dependent mechanical hysteresis:*** The occurrence of thermomechanical hysteresis is one of the most noticeable characteristics of the effect known as shape memory. This indicates that the forward and reverse transformations take place at temperatures that are distinct from one another. The hysteresis temperature is defined as the temperature difference that exists between the forward transformation (martensite-to-austenite) and the reverse transformation (austenite-to-martensite) [6].

  The hysteresis temperature has an effect on the shape memory qualities of the SMA. These features include the shape recovery speed as well as the stability and mechanical behavior of the SMA. Adjustments can be made to the composition, microstructure, and processing parameters of the SMA in order to get the desired effect.

In conclusion, the shape memory effect is a unique phenomenon demonstrated by SMAs, which allows them to restore their original shape after severe deformation. Shape memory alloys (SMAs) are examples of materials that exhibit this ability. Because of its one-of-a-kind behavior, a wide variety of sectors have been able to build ground-breaking applications. The application of SMA materials in functional composites and constructions is becoming increasingly viable thanks to ongoing research and technological breakthroughs in this field.

2.2.4 Super Elasticity

Another notable quality of shape memory alloys is their ability to exhibit a rubber-like behavior, which is also known as super elasticity (SMAs). It refers to the ability of shape memory alloys (SMAs) to withstand substantial elastic deformations and restore their original shape upon unloading. This is possible even after the SMA has been subjected to enormous strains that are in excess of the conventional elastic limit.

The martensitic phase of SMAs, which is the low-temperature phase, is where the majority of observations of super elastic behavior are made. The phenomena of super elasticity are linked to a transition known as stress-induced martensitic transformation, which takes place when mechanical stress is applied to a material. The following is an in-depth description of super elasticity in terms of SMAs:

- ***Behavior related to stress and strain:*** When a stress is applied to traditional materials, such as metals, that is greater than the material's yield point, the material goes through a process of permanent plastic deformation. SMAs, on the other hand, display a unique stress-strain behavior throughout the process of super elastic deformation.

  SMAs in their martensitic phase exhibit a nearly constant stress over a huge strain range when they are subjected to an external mechanical load. This is the case even if the strain range is quite broad. The super elastic plateau is the name given to this particular location. Because the tension in the plateau region is more or less constant, there is a significant amount of strain that may be recovered [7].
- ***Stress-induced martensitic transformation:*** The super elastic behavior of SMAs can be traced back to a stress-induced martensitic transformation that is reversible. When an external mechanical load is given to the SMA when it is in its martensitic phase, the stress causes the material to undergo a phase transformation from martensite to austenite. This is because austenite is a more stable phase than martensite.

  This change is caused by stress, and it takes place within the material as a result of the nucleation and proliferation of martensite variations. The reorganization of the crystal lattice on the level of the atoms makes it possible for the material to accommodate the stress that is applied.
- ***Storage and retrieval of elastic energy:*** During deformation in the super elastic region, the SMA stores elastic energy within its microstructure. This energy can then be recovered after deformation. This energy is retained in the reorientation and reformation of martensitic variations, which can be found in the material. After being unloaded, the elastic energy that was stored in the material is allowed to be released, which results in the material regaining its initial shape. The SMA is able to revert back to its initial

configuration with a high level of precision after undergoing the transformation from martensite to austenite.

- ***Dependence on temperature:*** The super elastic behavior of SMAs is impacted by the temperature at which the material is held. The martensite starts temperature (Ms) and the martensite finish temperature (Mf) are the critical temperatures that govern the martensitic transformation and the super elasticity zone (Mf) [8].

  "Hysteresis area" is a term that is frequently used to refer to the temperature range in which "super elastic behavior" can be observed. It is the temperature range between the melting point (Mf), where the change from austenite to martensite begins, and the austenite finish temperature (Af) (where the reverse transformation from martensite to austenite completes upon heating). The super elastic behavior and recovery properties of SMAs are both determined by the temperature at which the hysteresis point is reached, in addition to the stress–strain response.

In conclusion, the unique attribute of super elasticity that is possessed by SMAs permits enormous elastic deformations while yet allowing the material to restore to its original shape. Because of this feature, sophisticated applications have been developed for a variety of different industries, highlighting the adaptability and promise of SMAs in functional composites and structures.

2.3 Processing Techniques Used for Shape Memory Alloys

Techniques used to process shape memory alloys (SMAs) play a significant part in establishing the microstructure, phase transformation behavior, and mechanical properties of these materials. Casting, powder metallurgy, and thermomechanical processing are the three most frequent procedures utilized in the manufacturing of SMAs. Here is a description of each approach.

2.3.1 Casting

Casting is a conventional method for processing SMAs that involves the solidification of the alloy from a molten state into a specified shape.

This method involves pouring molten alloy into a mold to form the desired shape. The following procedures are included as part of the casting process for SMAs:

- ***Melting:*** The SMA alloy is heated to a high temperature until it enters a molten state, often in a controlled atmosphere or a vacuum. Once it has reached the molten state, the SMA alloy is melted.
- ***Molding:*** This process involves pouring the molten alloy into a mold or die chamber that already has the desired shape. Graphite or ceramic are two examples of refractory materials that are commonly used to construct the mold.
- ***Solidification:*** This step involves the cooling and solidification of the molten alloy within the mold, with the alloy taking on the shape of the mold cavity as it does so. To get the required microstructure and phase distribution, one must first carefully cool the material at controlled rates and under carefully monitored circumstances.
- ***Post-processing:*** After the solidification stage, the cast SMA is subjected to a variety of post-solidification operations, including heat treatment, hot/cold working, and surface treatments, in order to improve its physical characteristics.

Casting is a process that is both versatile and cost-effective when it comes to the production of SMAs with complex shapes. On the other hand, this could lead to microstructural flaws and make it more difficult to achieve fine-grained microstructures.

2.3.2 Powder Metallurgy

Powder metallurgy, often known as PM, is an additional typical processing technique used for SMAs. It is utilized most frequently for the production of components with regulated compositions and microstructures. The following procedures are included as part of the powder metallurgy process for SMAs:

- ***Preparation:*** The preparation of the powder involves combining the elemental powders of the SMA ingredients in the appropriate quantities. Methods such as mechanical alloying and gas atomization can be utilized in order to produce a powder composition that is consistent throughout.

- ***Consolidation:*** In this step, the powders are blended together before being compressed together under a great deal of pressure to create a green compact. The necessary shape is maintained by this compact, however it does not possess adequate strength.
- ***Sintering:*** This process involves exposing the green compact to increased temperatures that are lower than the alloy's melting point. The powders will bind together and undergo the desired phase transformation as well as microstructural alterations during the process is called sintering.
- ***Post-processing:*** After the sintering step, other operations like hot/cold working, heat treatment, or surface treatments can be applied to the SMA in order to further enhance its properties.

Powder metallurgy provides exceptional control over the composition of SMAs, as well as their porosity and microstructure. It makes it possible to produce components that are close to their net shape and have superior mechanical qualities.

2.3.3 Thermomechanical Processing

In thermomechanical processing, the microstructure and characteristics of SMAs are modified by combining mechanical deformation with heat treatment in order to get the desired results. The following procedures are included in this method, which is frequently utilized for NiTi-based SMAs:

- ***Hot working:*** In this process, the SMA is heated to a temperature that is higher than the temperature at which it recrystallizes, and then it is deformed mechanically using a process such as forging or rolling. The grain structure can be refined through hot working, which also helps to encourage the production of phases that are desirable and improves the workability of the material.
- ***Cold working:*** After the SMA has been subjected to hot working, it is allowed to cool and is then subjected to additional deformation at a temperature that is either at or below its recrystallization temperature. The material's capacity to withstand strain and grow in strength is enhanced by the application of cold working.

- ***Heat treatment:*** After the SMA has been deformed, it is put through a series of heat treatment operations, such as annealing or aging, in order to exert control over the phase change temperatures, the microstructure, and the mechanical properties. In order to achieve the desired characteristics, the process of heat treatment may involve procedures such as solution treatment, quenching, and aging.
- ***Post-processing:*** In order to get the correct surface qualities and shape for the finished product, additional post-processing operations such as machining, surface treatments, or coating may be necessary.

SMAs may have their microstructure, phase transformation temperatures, and mechanical properties precisely controlled through the use of thermomechanical processing. It is frequently employed in the production of components with a particular shape memory or with super elastic properties.

These many methods of processing each come with their own set of benefits and drawbacks. The processing method that is used is determined by a number of parameters, including the intended shape, the complexity of the component, the material composition, and the unique needs of the application [9].

2.4 Characterization Methods Employed to Evaluate the Microstructural and Mechanical Properties of SMAs

It is essential to characterize the microstructural and mechanical properties of shape memory alloys (SMAs) in order to gain a better understanding of their behavior and to enhance their effectiveness in the context of particular applications. SMAs are analyzed using a number of different characterization strategies, such as microscopy, X-ray diffraction, and mechanical testing, amongst others. Let us get into more depth about these different approaches.

2.4.1 Microscopy

Techniques using microscopy offer extremely helpful insights into the microstructure, grain size, phase distribution, and flaws of SMAs. Microscopy techniques commonly utilized for characterization of SMA include the following:

- ***Optical microscopy:*** Optical microscopy makes it possible to observe the microstructure and surface features of SMAs at lower magnifications than other types of microscopies. It is able to expose grain boundaries, phase borders, and any visible faults or inclusions in the material.
- ***Scanning electron microscopy (SEM):*** SEM is capable of producing high-resolution images of the SMA's surface as well as its cross-section. It makes it possible to conduct an in-depth investigation of the microstructure, as well as the phase morphology and grain boundaries. In order to conduct an analysis of the elemental composition, scanning electron microscopy (SEM) can be combined with energy-dispersive X-ray spectroscopy (EDS).
- ***Transmission electron microscopy (TEM):*** TEM or transmission electron microscopy is a technique that enables researchers to examine the internal microstructure of SMAs at the nanoscale. It is possible to obtain information about the crystal structure, phase distribution, and defects like as dislocations and stacking faults using this method [10].

2.4.2 X-Ray Diffraction (XRD)

X-ray diffraction is an effective method for evaluating the crystallographic phases, the behavior of phase change, and the texture of SMAs. The following are important XRD methods that are utilized for SMA characterization:

- ***Phase identification:*** The XRD can be used to help identify the crystallographic phases that are present in the SMA, such as austenite or martensite. Additionally, it is able to determine whether secondary phases or precipitates are present.
- ***Phase transformation analysis:*** XRD can be used to examine the behavior of phase transformation and determine the temperatures at which it occurs (Ms and Af). In order to get a better understanding of the transformation kinetics, heating and cooling cycles can be undertaken while keeping an eye on the changes in diffraction patterns.
- ***Analysis of texture:*** XRD makes it possible to assess the crystallographic texture or preferred orientation of grains in SMAs, both of which might have an effect on the materials' mechanical and functional properties [11].

2.4.3 Tests of the Material's Mechanical Characteristics and Behavior

Mechanical testing is absolutely necessary in order to characterize the mechanical properties and behavior of SMAs. The following are important mechanical tests that are used for characterization of SMA:

- ***Tensile testing:*** Tensile testing examines the stress-strain behavior, which is essential for designing SMA components. This behavior includes elastic modulus, yield strength, ultimate tensile strength, and elongation.
- ***Compression testing:*** Compression testing is used to evaluate the compressive properties of SMAs, including the compressive strength, strain, and energy absorption capacity of the material.
- ***Bending and flexural testing:*** These tests evaluate the bending behavior, flexural strength, and deformation capacity of SMAs. These tests are applicable for applications that involve bending or flexing.
- ***Fatigue testing:*** Fatigue testing evaluates the SMA's resistance to cyclic loads and determines its durability as well as its fatigue life.
- ***Testing for shape memory and super elastic behavior:*** These specialized tests evaluate the shape memory effect and the super elastic behavior of SMAs, including recovery stress, transformation temperatures, and hysteresis.
- ***Testing for shape memory and hysteresis:*** This test determines whether or not a SMA has a shape memory effect.

Along with other techniques like differential scanning calorimetry (DSC), thermal analysis, and electrical characterization, these methods of characterization provide a full understanding of the microstructural and mechanical properties of SMAs. They are helpful in guiding the design of SMA-based systems and structures for various applications and in optimizing those designs.

2.5 Applications of Shape Memory Alloys as Functional Composites

As a result of their one-of-a-kind qualities and capabilities, shape memory alloys, also known as SMAs, have recently attracted a lot of attention and been used in a wide variety of ways as functional composites in a variety of different industries. In the fields of aerospace, automotive, medicine, and consumer electronics, some prominent applications of SMAs are discussed in the following subsections.

2.5.1 Aerospace

- ***Actuators and control systems:*** SMAs are used in aerospace applications for actuation and control systems. SMAs improve an aircraft's overall performance and efficiency as well as their ability to regulate the aerodynamic surfaces of the aircraft precisely and adaptively. SMAs are utilized in the construction of deployable structures including as antennas, solar arrays, and morphing wingtips. They enable space applications and satellite systems to have compact storage, efficient deployment, and structural integrity [12].

2.5.2 Automotive

- ***Smart materials for automotive components:*** Smart material arrays (SMAs) are employed in automotive applications for a variety of components, including smart valves, engine mounts, and exhaust system elements. They make it possible for automobile systems to actively manage vibrations and dampen them as well as reduce noise and exhibit adaptive behavior [13].
- ***Suspension systems:*** SMAs are used in suspension systems to improve ride comfort, stability, and shock absorption. These benefits can be achieved through the use of shape memory springs and dampers. Spring and damper systems that are based on SMA technology have the ability to modify the stiffness and damping characteristics of their systems dependent on the driving conditions. This results in increased performance and safety.

2.5.3 Medical Applications

- ***Biomedical equipment:*** SMAs have many different uses in the medical industry. They exhibit super elastic behavior, which enables greater accommodation of physiological motions and minimises the effects of stress shielding when used in orthopedic implants such as bone plates, screws, and stents. These types of implants are employed because of their ability to hold their shape [14].

- ***Systems for the administration of drugs (SMAs):*** These are included into drug delivery systems so that medication can be administered in a regulated manner. In order to control the rates and patterns of medication release, microvalves and micro-actuators based on SMA can respond to external stimuli such as temperature and magnetic fields.
- ***Surgical instruments:*** Self-expanding stents, catheters, and retractors are some examples of the types of surgical instruments that make use of SMAs. They have the ability to remember their previous shape, which enables minimally invasive operations, improves mobility, and provides increased patient comfort.

2.5.4 Consumer Electronics

- ***Smartwatches and wearable gadgets:*** SMAs are integrated into smartwatches and wearable devices for the purpose of providing shape-changing interfaces, haptic feedback, and self-adjusting straps. They make it possible for dynamic and adaptive form factors to be implemented, which improves the overall user experience and comfort [15].
- ***Noise-canceling headphone and earphones:*** SMAs are used in noise-cancelling headphones and earphones to actively cancel or minimise unwanted sounds. This type of noise control is also known as "active noise control." They make it possible to change and adapt the acoustic qualities in order to improve the sound quality and lessen the amount of background noise.
- ***Micro-electro-mechanical systems (MEMS) actuators:*** SMAs are used in micro-electro-mechanical systems (MEMS) actuators for the precise positioning and control of small-scale devices. They make consumer electrical gadgets more responsive, more compact, and use less power. They also contribute to the shrinking of these products.

These examples demonstrate the adaptability of SMAs as functional composites and the great variety of uses that they can have. Because of its one-of-a-kind characteristics, such as the shape memory effect, super elasticity, and high actuation capabilities, shape memory alloys (SMAs) are extremely useful materials for the creation of creative solutions across a wide range of sectors.

2.6 Design Considerations and Challenges in Using SMAs for Specific Applications

When designing with shape memory alloys (SMAs), it is necessary to take into account a number of different elements and find solutions to a number of different issues in order to guarantee optimal performance and dependability for the intended applications. The following discussion presents a list of problems and design issues linked with the utilization of SMAs [16].

2.6.1 Compatibility with the Prerequisites of the Application

- ***Temperature range:*** The functionality of SMAs is dependent on certain transition temperatures (martensitic and austenitic), which are determined by the temperature range. It is the responsibility of the designers to guarantee that the operating temperature range of the SMA meets the requirements of the application.
- ***Strain and stress levels:*** The design should take into account the strain and stress levels that are anticipated during actuation or deformation. This will guarantee that the SMA is able to sustain and recover from the desired deformations without suffering any irreversible damage.
- ***Structural integrity and durability:*** This can be achieved by ensuring that the mechanical qualities of SMAs, such as stiffness, strength, and fatigue resistance, are compatible with the application that they are intended for.

2.6.2 Mechanisms of Actuation and Their Controls

- ***Actuation mechanism:*** The design ought to integrate appropriate mechanisms, such as shape memory springs, wires, or actuators, in order to make use of the exceptional features of SMAs in order to achieve the desired actuation or deformation.

- ***Control systems:*** Efficient control systems are required in order to successfully handle the activation, deactivation, and monitoring of SMAs. This is done in order to guarantee accurate and dependable performance. These control systems may include temperature regulation, mechanical activation, or responses to stimuli from the outside (such as electric or magnetic fields) [17].

2.7 The Choice of Materials and Their Compatibility

- ***Composition of the alloy:*** Choosing the correct composition of the SMA alloy is essential in order to achieve the required transformation temperatures, mechanical qualities, and environmental compatibility for a given application. Compositions of SMA that vary, such as nitinol (NiTi) or alloys based on copper, each have their own unique set of characteristics and behaviors.
- ***Compatibility with other materials:*** When SMAs are used in composite structures or in conjunction with other materials, it is essential to consider their compatibility and potential interactions, such as thermal expansion coefficients, bonding methods, or galvanic corrosion. This is because it is important to ensure that the SMAs and the other materials are compatible with one another.

2.7.1 Fatigue and Durability

- ***Fatigue life:*** Because of the cyclic loading or stress that they are subjected to, SMAs are susceptible to fatigue failure. To guarantee the SMA's durability and dependability in the face of repetitive actuation or deformation, its designers have to take into account a phenomenon known as fatigue behavior.
- ***Aging and deterioration:*** Over the course of their lifetime, some SMAs can experience aging or degradation, which can result in shifts in both their characteristics and their performance. It is imperative that these effects of aging be taken into account throughout the design phase in order to ensure that the required functionality is preserved over the product's anticipated lifetime of use [18].

2.7.2 Manufacturing and Processing

- ***Material processing:*** The manufacturing and processing procedures that are used should be compatible with SMAs in order to ensure that the necessary microstructure, phase transformation behavior, and mechanical properties are achieved. Casting, powder metallurgy, and thermomechanical processing are examples of processing technologies that need to be properly selected and carefully managed.
- ***Integration of components:*** In order to successfully integrate SMAs into complex systems or structures, considerable consideration of fabrication procedures, assembly methods, and compatibility with other components or materials may be required.

2.7.3 Expense and the Possibility of Commercialization

- ***Factors affecting price:*** The price of SMAs might change based on the alloy composition, the production processes used, and the supply chain. The designers need to evaluate the consequences of the cost and think about the trade-offs between the performance and the affordability of the product.
- ***Scalability in manufacturing:*** Ensuring that SMA production and fabrication methods can be scaled up to satisfy commercial demands and make the technology economically viable is one of the most important steps in developing SMA.

These design considerations and constraints underscore the necessity of having a full understanding of SMA behavior, compatibility, and manufacturing processes in order to assure successful integration and optimal performance in particular applications. It is absolutely necessary for scientists of materials, engineers, and application professionals to work together in order to handle these issues and come up with novel solutions utilizing SMAs.

2.8 Case Studies and Success Stories That Demonstrate the Practical Implementation of SMAs as Functional Composite Structures

In the following subsections are some examples of successful case studies and real-world applications of shape memory alloys (SMAs) in functional composite structures.

2.8.1 The Use of Smart Morphing Adaptors in Aerospace

The Adaptive Compliant Trailing Edge (ACTE) project at NASA integrated SMAs into the design of an aircraft wing. The SMA-actuated wing morphing technology enables shape changes to be made in real time, which improves both the aerodynamic performance and the fuel efficiency of the wing. The wing has SMA wires implanted in it, which allow it to alter shape and adapt to different flight circumstances. This helps to reduce drag and improves the aircraft's maneuverability [19].

2.8.2 Self-Deployable Space Structures

SMAs have been utilized in the creation of self-deployable space structures, such as solar arrays and antennae, which have been made possible thanks to their use. The Solar Array Systems (SAS) used by NASA make use of processes that are based on SMAs in order to achieve compact storage during launch and autonomous deployment in space. SMAs have the ability to exert a high force and remember their previous shape, which enables the reliable and controlled unfolding of massive structures in space [20].

2.8.3 Orthopedic Implants and Medical Devices

SMAs have fundamentally altered the landscape of the orthopedic implant and medical device industries. Nitinol-based SMA stents provide a remarkable illustration of this phenomenon. In the treatment of cardiovascular disorders including coronary artery blockages, these self-expanding stents are utilized as an effective therapy option. The SMA structure of the stent makes it possible for it to be administered using procedures that are minimally invasive. After delivery, the stent can expand to the required form, thereby restoring blood flow to the damaged area.

2.8.4 Adaptive Structures in the Automotive Sector

The adaptive structures that are based on SMA have been embraced by the automotive industry to improve vehicle performance and safety. SMAs are used, for instance, in BMW's Active Kidney Grille, which enables the opening and shutting of the grille to be controlled according to the amount of cooling that is required. When temperatures reach a certain threshold, the SMA wires become active and begin to regulate the airflow into the engine compartment. This results in an increase in the vehicle's overall efficiency.

2.8.5 Wearable Technologies and Intelligent Textiles

SMAs are finding more and more applications in the smart textiles and wearable technology industries. For instance, SMAs are utilized in fabric-based actuators in order to build clothes that are responsive and able to change shape. In order to enable applications such as adaptive clothing, smart compression garments, and dynamic fashion designs, researchers have developed SMA-embedded fabrics that are capable of either contracting or expanding in response to variations in temperature [21].

These case studies illustrate the practical application of SMAs in a variety of functional composite structures across a variety of sectors by presenting specific examples. SMAs are valuable materials for the development of creative and adaptable solutions because they offer unique qualities such as the shape memory effect, super elasticity, and high actuation capabilities. The success examples demonstrate the potential for SMAs to improve performance, efficiency, and the overall user experience across a wide variety of application types.

2.9 Current State of Research in Shape Memory Alloys and Potential Areas for Future Exploration

Research on shape memory alloys (SMAs) is ongoing, with the goal of bettering our knowledge of their characteristics, broadening the range of possible applications, and overcoming current difficulties. Here is a summary of the SMAs' research landscape today and some potential research directions for the future:

- Development and characterization of the material
  New SMA compositions and alloy systems are currently being developed in order to improve particular properties

like transformation temperatures, mechanical strength, and fatigue resistance. Alternative alloying components are being looked into by researchers, who are also examining how processing methods and microstructure affect SMA behavior. Phase transitions, defects, and microstructural changes in SMAs are investigated using cutting-edge characterization techniques such as electron microscopy, X-ray diffraction, and spectroscopy.

- Simulation and modeling:
  To comprehend the intricate behavior of SMAs, precise and predictive models must be created. In order to describe the nonlinear response, hysteresis, and energy dissipation in SMAs, constitutive models, including thermomechanical models, are currently being improved upon and validated. To bridge the gap between microstructural features and macroscopic behavior, multiscale modeling approaches incorporating atomistic, mesoscale, and continuum levels are being investigated.
- Modern manufacturing methods
  To enhance SMA processing, performance, and integration into complex structures, there is growing interest in researching advanced manufacturing techniques. The fabrication of intricate geometries and precise control over microstructures are made possible by the additive manufacturing (3D printing) of SMAs. To improve the characteristics of SMAs and make it possible for them to be incorporated into useful composite structures; other methods like selective laser melting, powder metallurgy, and shape casting are also being researched.
- Hybrid and functional composite materials
  Research is currently being done on the combination of SMAs and other materials to produce functional composites. In order to create materials with improved properties and multifunctionality, researchers are investigating the combination of SMAs with polymers, ceramics, and fiber-reinforced composites. Investigating the interface between the SMA and the matrix material, enhancing reinforcement techniques, and investigating the possibility of self-healing and self-sensing properties in SMA-based composites are just a few of the things covered in this.

- Upcoming applications
  SMAs have been successful in some fields, including aerospace and medicine, but there is still room for growth in other areas. Emerging fields like soft robotics, flexible electronics, energy harvesting, and shape-changing materials are being investigated by researchers. Exciting opportunities for exploration exist with the development of SMA-based artificial muscles, smart fabrics, and adaptable structures for next-generation technologies.
- Obstacles and restrictions
  SMA-related issues like fatigue, hysteresis, variability in properties, and the optimization of transformation temperatures for particular applications are still being researched. In order to better comprehend these difficulties and develop solutions, more research is required.

In conclusion, improvements in material development, modelling, manufacturing processes, and the investigation of novel applications characterize the current state of research in SMAs. Future research is anticipated to concentrate on improving manufacturing procedures, advancing modeling methodologies, and broadening the scope of applications for SMAs. Future research in this fascinating area will be fueled by continued interdisciplinary cooperation, integration with cutting-edge technologies, and a better understanding of SMA behavior.

2.10 Recent Advancements in the Development of Novel SMA Compositions, Processing, and Applications

The expansion of shape memory alloy (SMA) technology and its potential for various industries have been aided by recent developments in the development of novel SMA compositions, processing methods, and applications. A few noteworthy developments are discussed below.

2.10.1 Novel SMA Compositions

- ***SMAs with high temperature shape memory:*** Scientists have been investigating new SMA compositions with high temperature shape memory and super elastic behavior.

One example of this is the creation of the nickel–titanium–hafnium (Ni–Ti–Hf) and nickel–titanium–zirconium (Ni–Ti–Zr) alloys, which offer enhanced mechanical characteristics and shape memory performance at high temperatures.

- ***SMAs based on magnesium:*** Due to their light weight and potential uses in the aerospace and automotive industries, magnesium-based SMAs have drawn attention. In order to improve their mechanical properties, corrosion resistance, and shape memory behavior, recent research has concentrated on optimizing the composition and processing methods.

2.10.2 Advanced Processing Methods

- ***Additive manufacturing:*** Complex SMA structures have been created using additive manufacturing techniques with exact control over microstructures and geometries, such as selective laser melting and electron beam melting. This makes it possible to create specialized SMA components with better performance and integration capabilities, like actuators and sensors.
- ***Severe plastic deformation:*** In order to improve the mechanical properties and shape memory behavior of SMAs, the microstructure of these materials has been refined using severe plastic deformation techniques such as equal channel angular pressing (ECAP) and high-pressure torsion (HPT). These methods present a potential method for creating bulk SMAs with improved performance.

2.10.3 SMA Application Advancements

- ***Energy harvesting applications:*** SMAs have been investigated for energy harvesting applications, which involve converting mechanical vibrations or thermal gradients into electrical energy. The development of SMA-based energy harvesters with improved efficiency and integration into a variety of devices, such as wearable electronics and remote sensors, is a recent development.
- ***Soft robotics and biomedical applications:*** The fields of soft robotics and biomedical applications have become more popular as a result of SMAs. Artificial muscles based on SMA

are being developed by researchers to resemble the flexibility and range of motion of real muscles. These developments open the door to uses in prosthetics, assistive technology, and minimally invasive surgical instruments.
- ***Shape memory foams:*** Using SMAs, researchers have created shape memory foams that can be used for impact absorption, smart packaging, and cushioning materials. These foams could be useful in a variety of industries because they can deform significantly and then reform after activation.

2.10.4 Composite Hybrid Structures

- ***Hybrid composite structures:*** Recently, SMA integration into composite structures has been explored. SMAs are being used in the creation of hybrid composites, which have improved mechanical properties, shape memory functionality, and multi-functionality. These composites can also be made from other materials, such as polymers, metals, or carbon fibres. The aerospace, automotive, and civil engineering industries all use these hybrid structures.

These new developments show how SMA technology is constantly evolving, incorporating new compositions, processing methods, and applications. The use of novel processing techniques, the creation of advanced SMA compositions with enhanced properties, and the incorporation of SMAs into a variety of industries all contribute to the continued development and potential of SMA technology in the contemporary era.

2.11 Conclusions

In conclusion, our discussion on shape memory alloys (SMAs) has revealed their fascinating properties and wide range of applications. We have looked into their make-up, microstructure, and distinctive characteristics including the shape memory effect and super elasticity. SMA processing methods and characterization methods have been covered, with a focus on their utility in constructing complicated shapes and assessing properties.

SMAs offer improved performance and user experience and have found applications in the consumer electronics, automotive, aerospace, and medical fields. Design considerations and difficulties, such as hysteresis and fatigue, have been noted. However, continuous studies in cutting-edge manufacturing, innovative compositions, and developing industries like soft robotics and flexible electronics point to potential futures for SMAs.

SMAs show incredible promise as functional composites that can enhance a variety of systems. The future of SMAs relies on continued research, material advancements, and interdisciplinary collaborations.

References

1. Ares, A.E., *Shape-Memory Materials*, IntechOpen, London, 2018.
2. Xin, X., Liu, L., Liu, Y., Leng, J., Mechanical Models, Structures, and Applications of Shape-Memory Polymers and Their Composites. *Acta Mech. Solida Sin.*, 32, 535–565, 2019.
3. Lexcellent, C., *Shape-memory Alloys Handbook*. John Wiley & Sons, Inc. Hoboken, NJ., 2013,
4. Bhattacharya, K., *Microstructure of martensite: why it forms and how it gives rise to the shape-memory effect*, Oxford University Press, Oxford, 2003.
5. Chopade, J. *et al.*, Critical Review On Shape Memory Alloy. *J. Mater. Sci.*, 53, 12, 8765–8791, 2018.
6. Jani, J.M., Leary, M., Subic, A., Gibson, M.A., A review of shape memory alloy research, applications and opportunities. *Mater. Des.* (1980-2015), 56, 1078–1113, 2014.
7. Wayman, C.M., Shape memory and related phenomena. *Prog. Mater Sci.*, 36, 203–224, 1992.
8. Lobo, P.S., Almeida, J., Guerreiro, L., Shape Memory Alloys Behaviour: A Review. *Procedia Eng.*, 114, 776–783, 2015.
9. Yan, J.W. *et al.*, Texture Development in the Ni47Ti44Nb9 Shape Memory Alloy. *Metall. Mater. Trans. A*, 44, 3, 1421–1430, 2012.
10. Tiyyagura, M., Transmission electron microscopy studies in shape memory alloys, 2023.
11. Öner, E. *et al.*, Influence of Heat Treatment on Cu-Al-X Shape Memory Alloys. *J. Phys.Chem. Function. Mater.*, 5, 2, 45–58, 2022.
12. Duerig, T.W., Melton, K.N., Stöckel, D., Wayman, C.M. (Eds.), Shape Memory Alloy Applications In Space Systems, in: *Engineering Aspects of Shape Memory Alloys*, Butterworth-Heinemann, 1990.
13. Lagoudas, D.C., *Shape memory alloys: modeling and engineering applications*, Springer, New York, 2008.
14. Lecce, L. and Concilio, A., *Shape memory alloy engineering: for aerospace, structural and biomedical applications*, Butterworth-Heinemann, Oxford, 2015.
15. Cai, S., Xu, X., Yang, W., Chen, J., Fang, X., Materials and Designs for Wearable Photodetectors. *Adv. Mater.*, 31, 1808138, 2019.

16. Dye, S. and Ruth, D., *Shape Memory Alloy Actuators: Design and Control.* Wiley, Hoboken, NJ, 2020.
17. Liang, C. and Rogers, C.A., Design of Shape Memory Alloy Actuators. *J. Mech. Des.*, 114, 223–230, June 1992.
18. Le Gac, P.Y. *et al.*, Evaluation of Long-Term Behavior of Polymers. *Oil Gas Sci. Technol.*, 70, 1, 123–135, 2014.
19. Hartl, D.J. and Lagoudas, D.C., Aerospace applications of shape memory alloys. *Proc. Inst. Mech. Eng. Part G J. Aerosp. Eng.*, 221, 535–552, 2007.
20. Li, F., Liu, Y., Leng, J., Progress of shape memory polymers and their composites in aerospace applications. *Smart Mater. Struct.*, 28, 103003, September 2019.
21. Miyazaki, S. *et al.*, *Thin Film Shape Memory Alloys: Fundamentals and Device Applications*, Cambridge University Press, Cambridge, 2009.

3

Characterization and Testing of Smart Functional Composites

Ranjita Swain[1], Sunita Routray[2*] and Rudra Narayan Mohapatro[1]

[1]Department of Chemical Engineering, C.V. Raman Global University, Bhubaneswar, Odisha, India
[2]Department of Mechanical Engineering, Centurion University of Technology and Management, Jatni, Odisha, India

Abstract

This chapter explores the cutting-edge advancements in smart functional composites, materials that exhibit enhanced properties and adaptive behaviors in response to external stimuli. It begins with a foundational overview of the concept of smart composites, which integrate functional components such as sensors, actuators, and responsive polymers into traditional composite matrices. These materials can adapt to changes in their environment, offering benefits like self-healing, damage detection, shape memory, and environmental responsiveness. The chapter covers key types of smart composites, including those embedded with piezoelectric, thermochromic, and magnetostrictive elements, and discusses their unique capabilities in enhancing the performance of structural materials. Furthermore, the chapter investigates the challenges involved in developing these materials, such as material compatibility, processing techniques, and durability in real-world applications. Key industries such as aerospace, automotive, and healthcare are highlighted, demonstrating the diverse applications and transformative potential of smart functional composites. The chapter concludes with a look toward the future, identifying emerging trends, including the use of sustainable materials and the incorporation of artificial intelligence (AI) for real-time monitoring and adaptive responses.

***Keywords*:** Composites, characterization, radiation, healthcare, AI, applications

**Corresponding author*: sunitaroutray77@gmail.com

Sandip Kunar, Pranav Charkha, Santosh Jaju and Harish Tiwari (eds.) Functional Composites: Role in Modern Engineering, (49–70) © 2025 Scrivener Publishing LLC

3.1 Introduction

Smart functional composites are used as advanced materials benefit for traditional composites with the ability to respond intelligently to external stimuli such as electric or magnetic fields, chemical environments, temperature, pressure, and light. These materials are adopted by industries to actuate, sense, or transform their properties based on changing its conditions. These materials have a wide number of advantages like ideal formation for different applications including healthcare, aerospace, electronics, automotive, and environmental monitoring. The characterization of these composite materials is required for understanding their performance, behavior, and potential applications. A material's composition, structure, and functionality can create vital roles for different properties like structural, thermal, mechanical, electrical, and other properties. The testing of functional composites is an analytical process for evaluating the suitability, reliability, and performance for specific applications. These materials have multiple applications like structural health monitoring, aerospace, electronics, and biomedical devices.

It ensures that the composite comes across the required standards for performance, durability, and safety.

3.1.1 Shape Memory Composites (SMPs)

The materials recover their original shape after being contorted in response to specific external stimuli such as heat, light, or an electric field. They combine the versatility of shape memory materials with the enhanced mechanical, thermal, or electrical properties of composite materials. The shape memory composites can recover their original shape when exposed to specific stimuli such as heat, light, or an electric field. The example of SMPs is reinforced with their fibers. Shape memory polymers: Shape memory polymers (SMPs) processes can apply in aerospace structures, medical devices (e.g., stents), and deployable components [1].

3.1.2 Self-Healing Composites

Engineers can autonomously repair damages such as cracks, scratches, and other forms of degradation with minimal external intervention in the self-healing process and extending the lifespan of the materials. Some natural healing processes have also been adopted for living organisms, these materials also have been designed particularly in applications where regular maintenance or replacement is challenging or costly. The applications of self-healing composites are structural health monitoring, coatings, automotive parts, etc. [2].

3.1.3 Piezoelectric Composites

Piezoelectric composites are advanced materials that combine piezoelectric properties with the structural benefits of composite materials. These materials generate an electrical charge in response to direct piezoelectric effect or deform when exposed to an electric field (inverse piezoelectric effect). These materials generate an electrical charge in response to mechanical stress or, conversely, deform under an electric field, typically made by embedding piezoelectric ceramics or polymers into a polymer matrix. The applications of piezoelectric composites are in sensors, actuators, and energy harvesting devices [3].

3.1.4 Magnetostrictive Composites

Magnetostrictive composites are advanced materials that exhibit a change in shape, when subjected to a magnetic field (magnetostrictive effect). Conversely, they can also generate a magnetic field in response to mechanical stress (inverse magnetostrictive effect). These composites combine magnetostrictive materials, such as magnetostrictive alloys or ceramics, with flexible or durable matrices to enhance mechanical, thermal, and magnetic properties.

The applications of magnetostrictive composites are vibration control, actuators, and magnetomechanical sensors [4].

3.1.5 Thermoelectric Composites

Thermoelectric composite is one of the advanced materials used to covert temperature gradients into electric energy which is known as Seebeck effect and, in another way, generates a temperature difference when the electrical current flows through them, which is known as Peltier effect. When both thermoelectric materials with other components combine, then those composites achieve the performance, flexibility, and adaptability for a wide range of applications. These innovations promise to unlock the full potential of thermoelectric composites in energy conversion, sustainable power generation, and advanced thermal management systems. Hybrid composites of thermoelectric materials (e.g., bismuth telluride) are submerged in polymer or ceramic matrices. The applications of thermoelectric composites are energy harvesting, cooling systems, and wearable electronics [5].

3.1.6 Conductive Composites

Conductive composite material can be designed to get the lightweight properties, mechanical strength, and versatility of composite material with electrical and thermal conductivity. Conductive materials such as metals, carbon-based materials (e.g., graphene, carbon nanotubes), or conductive polymers, such as insulating matrix like polymers, ceramics, or resins, can be used to get the composite material. These conductive composite materials are widely used in industries, where multi-operational properties are required, like electronics, aerospace, and automotive. This type of material is essential for the phenomenon of lightweight, high-performance, and multifunctional materials, paving the way for innovations in latest electronics and advanced engineering solutions. The applications of conductive composite are flexible electronics, electromagnetic interference (EMI) shielding, and energy storage devices [6].

3.1.7 Light-Responsive Composites

Light-responsive composites are materials that pass through shape memory and shape healing capacity. The composite material is also designed for changing their properties in response to exposure to ultraviolet (UV), visible, or infrared (IR) light. These composite materials combine with light-sensitive materials, such as photoactive compounds, and with traditional materials such as polymer type, ceramic types, or metal types of material. It was also observed to show different behaviors, change of color, bending process, or triggering of chemical reactions. They are widely used in smart windows, sensors, and optical devices. It is also used in smart material and adaptive systems [7].

3.1.8 Bio-Inspired Composites

The bio-inspired composite type of material is also known as biomimetic composites, where these composites integrate principles from biological systems to create materials with enhanced mechanical, thermal, or environmental performance. This composite has remarkable properties such as self-healing, lightweight strength, flexibility, and environmental sustainability. Bio-inspired composite materials can be used in biomedical devices, soft robotics, and green technologies [8].

3.1.9 Multi-Functional Composites

Multi-functional material is one of the advanced materials engineered to perform multiple distinct functions simultaneously. These materials have different properties and are typically found in separate materials, which can adapt for multiple tasks. These materials can be used in industries like aerospace, automotive, electronics, and healthcare, where multifunctionality, efficiency, and lightweight materials are crucial. It is also used in advanced aerospace systems, smart infrastructure, and integrated sensors. The multiformity of smart functional composites offers suitable solutions to meet the different challenges of modern engineering and technology [9].

The global composite market size was calculated at USD 111.89 billion in 2023 and is expected to reach around USD 191.36 billion by 2032, growing at a significant CAGR of 6.1% from 2023 to 2032 (Figure 3.1). Some of the key factors of composite market growth include the increasing demand for lightweight materials in defense, automotive, and aerospace sectors as well as the rising need for chemical and corrosion-resistant materials in the pipe, tank, and construction industries. Furthermore, the growing requirement for flame-retardant and electrically resistant materials in electronics and electrical applications further boosts market expansion.

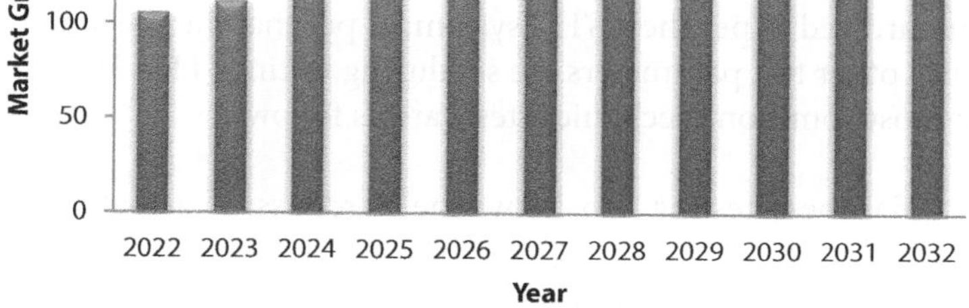

Figure 3.1 Expected global market growth in dollar from 2022 to 2032 [10].

Emerging trends, such as the development of cost-effective carbon fibers, faster-curing resin systems, and high-performance glass fiber, are also positively shaping the dynamics of the composite market.

3.2 Mechanical Characterization

3.2.1 Overview of Mechanical Testing Methods

Mechanical testing is a vital part of the manufacturing industries. It is done to assess the suitability of the products to the service conditions. Unless the product meets the service conditions, it is not accepted by the customers. Most of the manufacturing industries have their own testing laboratories, where the products are tested under the service conditions to which it will be put. For manufacturing industries where a testing facility is not available, the products are sent to external testing laboratories. Mechanical testing is done to assure the safety of the final products or any structure. The environmental conditions are also important; hence, testing must be conducted under similar conditions, which the product will face during its service period. Mechanical testing is of two types: destructive and nondestructive. The parameters that can be studied from destructive mechanical testing are yield strength, ultimate tensile strength, failure strength, percentage elongation, percentage reduction in area, compression strength, creep strength, fatigue strength, hardness, impact strength, flexural strength, etc. In nondestructive testing, the product remains as it is, and it is particularly used to detect cracks present on the surface and inside the product. The strength as well as failure analysis are both necessary for a product before putting it in service condition. Another parameter, the factor of safety, is also important for any product or structure in its service condition. The factor of safety is a fundamental concept in design and defined as the ratio of the maximum strength of a material to the working strength applied to it. Generally, for safety purposes, a lower amount of load is applied to the structures than the maximum load it can sustain to avoid failure. The tests are standardized as per the ASTM system; as per that the specimen dimensions and other test parameters are set during testing [11–13].

The most common mechanical tests are as follows:

- Hardness testing—to know the hardness of a material. It includes:
  - ➢ Vickers hardness test (HV): It is extremely useful in testing the hardness of objects, such as foils.

- ➢ Brinell hardness test (HB): It is suitable for measuring the hardness of hard materials that need more force for penetrating the surface.
- ➢ Knoop hardness test (HK): This is useful for brittle materials such as ceramics due to their elongated indenter.
- ➢ Janka hardness test: This method is useful for measuring the resistance of wood to denting and wear.
- ➢ Meyer hardness test.
- ➢ Rockwell hardness test (HR): This is most commonly used. It provides a quick and accurate measure of hardness on almost all metals and some plastics.
- ➢ Shore durometer hardness: It is used for measuring the hardness of rubbers and plastics.
- ➢ Barcol hardness test: This testing method is commonly used for composite materials such as reinforced thermosetting resins.

- Tensile testing is used to assess the mechanical strength of a material. It is used to obtain the engineering stress–strain curve for a material. Some of the important parameters can be calculated from this curve, such as Young's modulus, yield or proof stress, tensile stress, fracture stress, % elongation, and % reduction in area. The data obtained from this testing is very much useful for design purposes.
- Impact testing is used to determine the toughness of a material.
    - ➢ Izod test: This impact test is used to determine the toughness of brittle materials like plastics.
    - ➢ Charpy test: This is a type of impact testing used to determine the toughness of ductile materials like metals.

- Creep testing is used to determine a material's long-term strength and heat resistance to predict the lifespan under certain operating conditions.
- Fatigue testing is used to study the behavior of materials under cyclic loading.
- Non-destructive testing (NDT) is done without damaging the parts to detect cracks at the surface as well as in depth. There are different types of NDT methods, which are presented below.

- ➢ Radiographic testing
- ➢ Ultrasonic testing
- ➢ Liquid penetrant testing
- ➢ Magnetic particle testing
- ➢ Eddy current testing
- ➢ Magnetic flux leakage testing
- ➢ Thermal/infrared testing
- ➢ Acoustic emission testing
- ➢ Ground penetrating radar
- ➢ Guided wave testing
- ➢ Microwave testing
- ➢ Laser testing
- ➢ Leak testing
- ➢ Vibration analysis

3.2.2 Discussion of Mechanical Properties

Different mechanical tests are conducted to evaluate a material's strength parameters. These parameters are very useful for the design engineers to put the materials in proper service conditions as per the service requirements. All decisions related to material selection, design, and manufacturing taken by engineers are based on the results obtained from mechanical testing [14].

The properties of materials are as follows:

- Elasticity: The material comes back to its original shape and size when the applied force is removed.
- Plasticity: The deformation takes place in the material due to application of an external force is permanent and nonrecoverable.
- Stress: It is a restoring force in the opposite direction to the deforming force. This restoring force per unit area is termed as stress. The magnitude of the stress = F/A.
- Strain: Due to application of stress to a material, it elongates. Strain is defined as the ratio of change in dimension to the original dimension.
- Hooke's Law: It is the relationship between stress and strain, determined from the stress–strain diagram of a material. According to Hooke's law, stress is directly proportional to strain.
- Stress–strain curve: This curve provides the Young's modulus, yield stress, ultimate stress, fracture stress, % elongation,

and % reduction values of a material, which are very useful for designing purpose.

- Young's modulus: It defines the relationship between stress and strain for a material. Every material has a different Young's modulus value.
- Toughness: It is the amount of energy absorbed by any material undergoing plastic deformation until fracture.
- Ductility: It is the property by which a material can be drawn into wires. The material should be a ductile material to achieve high elongation.
- Malleability: It is the property by which a material can be drawn into sheets. The material should be a ductile material to achieve large plastic deformation.
- Hardness: It is the property of a material which gives it resistance to any kind of scratch, indentation, or rebound. Accordingly, the hardness value is named so; scratch hardness, indentation hardness, and rebound hardness.
- Brittleness: The material which breaks into pieces without any or very little plastic deformation upon the application of load is termed as a brittle material. Brittleness is opposite of the property ductility.

3.2.3 Mechanical Characterization of Smart Functional Composites

Smart functional composites, also regarded as intelligent composites, have the ability toward environmental stimuli in a predictable manner. These composites are prepared by combining different materials with variable chemical or physical properties. These composites are useful for a variety of applications such as flexible electronic devices, self-healing and super hydrophobic coatings, shape memory materials, chemical and strain sensing, energy harvesting and storage, actuators, switches, robots, artificial muscles, etc. These are capable of responding to various factors like temperature, pressure, electric fields, magnetic fields, chemical fields, mechanical stress, and radiation. Some challenges are faced during manufacturing of smart functional composites. Meticulously, it was tried to achieve a balance between the mechanical, electrical, and thermal properties of the composite material keeping the cost-effectiveness in mind [15].

3.2.4 Electrical Characterization of Smart Functional Composites

Electrical testing methods are generally meant for measuring the resistance, conductivity, or charge storage capacity either on the surface of the material or through it. Properties like dielectric strength, volume, and surface resistivity are very important to assess a material.

3.2.4.1 *Dielectric Strength*

Dielectric strength is the property which indicates the breakdown strength of a material. After the material reaches the breakdown point, it is no longer an insulator and electrical current breaks through the material. The procedure is to test the material in a transformer oil test kit with 220 V, 50 Hz, single-phase AC supply. The positive and negative terminals of the instrument get connected at both ends of the material through a highly conducting copper metal, and then the voltage is gradually increased until the voltage increment breaks and begins to drop, which indicates the breakdown voltage of the sample.

The dielectric strength of the insulator is calculated by using the formula:

dielectric strength = breakdown voltage (kV) / thickness of sample (mm)

3.2.4.2 *Insulation Resistance*

The insulation resistance test is carried out to measure the resistance between any two points separated by an electrical insulation. It is done by using a megger insulation tester. The tester gives a digital value of the insulation resistance. A product is tested at its early life and in between service application to check its efficiency.

3.2.4.3 *Volume Resistivity and Surface Resistivity*

Surface resistivity is measured in terms of resistance to leakage current at the surface of any material, whereas volume resistivity is the resistance to leakage current inside the body of the material. The ratio of surface to volume resistivity indicates the conductive nature of the material. The lower the ratio, the less conductive the material is.

3.2.5 Types of Smart Functional Composites

The smart functional composites can be divided into four categories depending upon their functions in different areas.

The smart functional composites are as follows:

- Structural smart composites
- Composites for actuation
- Novel functional composites
- Nanocomposites that are enablers of novel functions

3.2.5.1 Structural Smart Composites

Structural smart composite materials are capable of sensing stress, strain, fatigue, and damage of structures. These are also capable of monitoring the health conditions of critical structures like the blades of a wind turbine, underground pipes, and long-span bridges. Smart materials are being embedded into the structural material to make a compact design. Sensors are also incorporated into the design to make the structure more reliable. Piezoelectric fiber optic sensor is widely adopted for this purpose and gives satisfactory results.

To monitor the conditions of the structures, fiber optic sensors are embedded in carbon-fiber-reinforced polymer composites. These sensors generate the conditions of the structures periodically during their service condition.

In cement-based piezoelectric ceramic composites, piezoelectric fiber optic sensors are also embedded. Whenever it experiences any mechanical stress, the sensor generates an electric charge. This property of the piezoelectric sensors is reversible, meaning that the material experiences stress when coming across an electric field. This property allows smart composites to sense and also to actuate. Lead zirconate titanate piezoelectric ceramic powders and white cement matrix are used to prepare cement-based piezoelectric composites [16].

3.2.5.2 Composites for Actuation

Shape memory alloy (SMA) materials remember their original shape as the name implies. They are also called memory metal, memory alloy, smart metal, smart alloy, and muscle wire. SMA has the property by which it can take significant pseudo-elastic deformations. Upon exposure to stimuli like temperature, stress, or strain, it again returns to its original shape and size. Some of the examples of SMAs are copper–zinc–aluminum–nickel, copper–aluminum–nickel, and nickel–titanium (NiTi) alloys [17].

3.2.5.3 *Novel Functional Composites*

Artificial skins prepared with the combination of stretchability and sensing capability are popular in use. These skins are soft and stretchable with electronic sensing components embedded into it. Conductive elastomeric composites, with carbon nanotubes incorporated into it, are stretchable, transparent, and ultrasensitive. These artificial skins can sense touch, temperature, humidity, etc. The color of artificial skins can also be changed by embedding organic electrochromic devices.

3.2.5.4 *Nanocomposites for Novel Functions*

Many smart functions can be enabled by incorporating nanoparticles into a composite. The synthesized Fe_3O_4 multi-walled carbon nanotubes are used to fabricate microwave absorber materials. Nanocapsules of functional substances which are applied to the fabrication are considered as smart materials.

3.3 Thermal Characterization of Smart Functional Composites

The advanced materials which are integrated into two or more materials such as ceramics, metals, polymers, carbon-based materials, etc., enhance the properties of the newly created material. These smart functional composites have the capability to stimulate properties such as temperature, light, electrical fields, or mechanical forces. The combination of materials in a conventional way improves the strength and toughness, whereas functional composites have the advantage of flexibility and are capable of self-sensing, which sanction their characteristics in the real world [33].

Smart composites functionalize to get benefits in the biological systems that interact with the environment. Smart composites have characteristics such as self-healing, shape memory, adaptive stiffness, or capability to generate the energy (such as piezoelectric composites). The abovementioned characteristics create a great variety of applications which include automotive, medical devices, renewable energy systems, and aerospace.

These composites are engineered to possess "smart" functionalities that enable them to interact with their environment in a way that mimics biological systems, offering a range of benefits. For example, smart composites may exhibit features like self-healing, shape memory, adaptive stiffness, or the ability to generate energy (such as piezoelectric composites). These properties make them highly suitable for a wide variety of applications, including aerospace, automotive, medical devices, robotics, and renewable energy systems.

The most important advantage of this composites is to increase the performance and longevity of structures to reduce external involvement or maintenance. The material's life cycle will be extended by the self-healing process of cracks automatically within a short span of time. The composites of shape memory give rise to new opportunities in the area of actuators or adaptive surfaces. Nowadays the advancements are due to novelties in the field of nanotechnology, material science, and manufacturing techniques. The smart functional composites are controlled to get a revolution of industries which require materials that offer mechanical strength and also respond to the environment.

Thermal characterization is a vital role to evaluate the presentation and productivity of the functional composites. Thermal behavior is necessary to ensure the consistency, performance, and efficiency of the smart materials for application in many sectors such as healthcare, aerospace, automotive, and energy.

3.3.1 Thermal Behavior of Smart Functional Composites

This behavior of smart functional composites states the response of the materials with change in temperature. Temperature affects physical and chemical properties such as shape, conductivity, and structural integrity.

The thermal conductivity of the composites is going to bear the temperature impacts on the performance in the application of thermal management. The composite materials change their shape with temperature such that applications must be fit for the change of shape without any failure. In the glass transition temperature, the materials transit from a rigid to a flexible state that is important for polymer-based composites. The shape, size, and mechanical properties of the smart composites change with temperature response as found in shape memory polymers or shape memory alloys fixed in the matrix of the composites.

3.3.1.1 Thermal Characterization Techniques

There are different types of techniques being used to measure the thermal properties of smart functional composites. Differential scanning calorimetry is used to measure the rate of heat transfer in the phase transition such as glass transition, crystallization, and melting. The thermal stability of smart composites is applied to their temperature change. The authors used DSC to investigate the thermal behavior of shape memory polymer composites, showing distinct transitions that correspond to shape recovery at elevated temperatures. Thermogravimetric analysis (TGA) is used to measure the weight loss of the material with change in temperature. It is used to measure thermal stability and degradation temperatures [26]. The studies were carried out for the thermal degradation of self-healing composites using TGA. The result is indicated to show a strong correlation between the encapsulated process and thermal stability. Laser flash analysis (LFA) is a method dealing with the thermal diffusivity of materials and the low thermal conductivity of the composites which are applied in the insulations [21]. The authors used LFA to characterize the thermal conductivity of a grapheme-based polymer composite signifying the improvement in the efficiency of heat transfer rate. Heating and cooling cycles help in thermal stability and smart composites. This process undergoes a phase transition or shape memory effect that is critical for materials [18]. As per previous studies on thermal cycling on shape memory, composite materials reveal a shape recovery retention capability even after 1,000 cycles. The non-destructive characterization method is applied to measure the temperature distribution on the surface of the functional composite during heating [24]. The authors applied infrared thermography to assess thermal performance and the behavior of self-healing composites under thermal stress [20].

3.3.1.2 Thermal Properties of Key Smart Functional Composites

Shape memory composite (SMC) materials show reversible changes when open to a specific temperature. The glass transition temperature (T_g) and the activation temperature required for shape recovery are involved in the thermal characterization of SMEs. Using the thermal behavior of a shape memory polymer composite, it was found that the material exhibited an efficient shape recovery process when subjected to heating above its Tg. The self-healing composite characterization is required to measure the activation energy [19]. TGA and DSC were used to assess the thermal stability and healing behavior of self-healing composites, revealing that the encapsulated healing agents were activated at around 100°C [22]. Wang *et al.* (2023) observed color changes under establishment through reversibility at 45°C to save energy whenever he studied a thermochromic composite material. The functional smart materials change the color and transparency according to light and temperature. The color of the materials changes with the temperature and becomes stable after several cycles of experiment using a thermal characterization system. The sensors or energy harvesting devices show the critical behavior using temperature. The reliability of the material is ensured by the effect of temperature on thermal expansion and electrical conductivity [23].

The author conducted a study on piezoelectric composites for energy harvesting, showing that temperature variations influenced their output performance, requiring careful thermal characterization to optimize their design [25].

3.3.1.3 Challenges and Future Directions

There are also many challenges found in the thermal characterization techniques for optimizing and developing smart functional materials as this technique is well established. The combination of the different materials with different thermal characterizations leads to a thermal gap that results in stress and potential failure. There is another challenge in the shape memory and self-healing materials of functional smart composites. The functional smart composites show characterizations such as thermal, mechanical, and electrical properties simultaneously that balance the characterizations for the applications as a challenge. The advanced thermal characterization techniques may be adopted for future research programs.

Thermal characterization technique is very necessary to design and apply in smart functional composites. Techniques like DSC, TGA, LFA, and infrared thermography include valuable visions into the thermal behavior of the functional composite materials. The advancement in thermal characterization as the research is in progress enhances the materials applied in the area of aerospace, automotive, and energy systems.

3.4 Environmental and Durability Testing of Functional Smart Materials

Resistance to environmental factors like humidity, UV radiation, temperature, mechanical stresses, corrosive elements, and fluctuations are characterized by environmental and durability testing of smart functional composites. The characterization study includes performance efficiency with respect to the time, specifically aging, cyclic charging, and disclosure to the exact operating conditions.

There are five types of environmental characterization of functional smart materials, namely: (i) thermal cycling and temperature testing, (ii) moisture and humidity testing, (iii) UV and radiation exposure, (iv) mechanical and vibration testing, and (v) electromagnetic interference (EMI) testing.

3.4.1 Thermal Cycling and Temperature Testing

The efficiency performance of functional smart composite materials characterized through thermal cycling is critical as these materials are involved in the variation in temperature during their life period. Due to temperature fluctuation, smart composites used in aerospace and automotive sectors are subjected to the thermal stress. Repetition of heating and cooling cycles for thermal cycle test of the materials increases the real-world condition and also assesses the degradation of the material, performance loss, and delamination [27].

It is demonstrated that the thermochemical characteristics of smart functional composites were degraded under precisions of thermal cycling specifically for the low-thermal-stability polymers [33]. The carbon nanotube prepared from smart functional nanomaterials can diminish these effects; the abovementioned overall effectiveness of this composite can be validated through continuous thermal testing.

3.4.2 Moisture and Humidity Testing

Humidity and absorption of moisture content, properties which affect the characterization of the mechanical and electrical properties of smart composites, are the ones used to measure composites that are integrated with polymer matrices. The electronic or piezoelectric sensors from the composites are highly susceptible to the moisture content which degrades the material, leading to corrosion, delamination, and swelling [30]. Durability and functionality are evaluated by testing of the moisture content effect after implication of salt spray, immersion tests, and controlled-humidity chamber. The author studied the effects of the absorption of moisture content in measuring the electrical conductivity of the smart composites and

CNTs. It is also observed that the humidity content decreases the conductivity because of the weakening of the interface in between the matrix and CNTs [34].

3.4.3 UV and Radiation Exposure

The mechanical and functional characteristics of the smart composites with organic polymer matrices are degraded by UV radiation and high-energy radiation. UV exposure can cause polymer chain scission, leading to reduced strength and flexibility. The embedded electronic sensors or energy-harvesting components may face efficiency decreases due to radiation [29].

The effects of UV exposure that has accelerated weathering tests such as QUV (UV exposure under controlled temperature and humidity conditions) are commonly employed. Recent studies carried out by authors for incorporating UV stabilizers, such as nano-TiO_2, into smart composites can significantly enhance their UV resistance and improve the material's long-term durability [35].

3.4.4 Mechanical and Vibration Testing

Under dynamic loading, mechanical and vibration testing are vital for evaluating the performance of smart functional composites under dynamic loading conditions. Composites can withstand cycling loading, impact stress, and vibration that is being used in aerospace structure. This is particularly important for smart composites with integrated sensors or actuators, as mechanical stresses can disrupt their functionality [31]. Vibration testing on composites to resonant frequencies and high-g loads is used to measure the fatigue behavior and the potential failure of embedded functional elements. Dynamic mechanical analysis (DMA) and fatigue testing under cyclic loading are standard techniques used to assess both the structural integrity and functional performance over time [36].

3.4.5 Electromagnetic Interference (EMI) Testing

Smart composites, especially those used in electronics or energy harvesting applications, often include electronic components that are sensitive to electromagnetic interference (EMI). Therefore, it is critical to assess how these composites perform when exposed to high-intensity electromagnetic fields or radiation. Testing typically involves subjecting composites to varying levels of electromagnetic radiation to determine the degradation of their electrical properties [33]. In the recent trend, the new material EMI used for the smart composites has a potential in optimizing the conductivity nanomaterial composites that give less impact on the mechanical and functional reliability while exposed to electromagnetic fields.

3.5 Durability Testing Methodologies for Smart Functional Composites

3.5.1 Accelerated Aging Testing

The accelerated aging tests are used to stimulate the durable environmental exposure in a short duration by using different conditions at extreme points such as high temperature, humidity, and UV radiation. These accelerated aging tests are designed to simulate long-term environmental exposure over a short period using extreme environmental conditions such as high temperatures, humidity, and UV radiation. These tests allow for the rapid prediction of the longevity of smart composites, especially for those used in outdoor or aerospace applications where exposure to various environmental elements is unavoidable [29].

For example, the authors used accelerated aging tests to investigate the long-term stability of piezoelectric smart composites used in structural health monitoring systems. They found that while the composites retained their piezoelectric properties over short periods of exposure, prolonged exposure to UV light and moisture significantly diminished the material's sensor sensitivity [32].

3.5.2 Self-Healing and Damage Detection

Smart composites often feature self-healing capabilities, where embedded healing agents or microcapsules can repair a damage when the composite is exposed to stress or mechanical deformation. Durability testing of such composites focuses not only on the material's resistance to environmental stresses but also on the effectiveness of the self-healing process. Techniques such as fracture toughness testing, damage monitoring, and real-time sensing are employed to assess the healing efficiency and long-term durability of these materials under cyclic loading and environmental exposure [28].

3.5.3 Field Testing and Real-World Simulations

While laboratory-based testing is crucial, field testing under real-world environmental conditions is essential for understanding the true durability and performance of smart composites. Field testing provides insights into how composites behave under prolonged exposure to fluctuating weather conditions, operational stresses, and unexpected environmental factors [27].

The new study conducted field testing on smart composites used for wind turbine blades and found that these materials could maintain their structural integrity and sensor functionalities even after 2 years of exposure to outdoor conditions [36].

3.6 Recent Advances in Smart Functional Composite Testing

3.6.1 Integration of Smart Sensors for Real-Time Monitoring

The integration of sensors into smart composites allows for real-time monitoring of material degradation during environmental testing. The use of strain gauges, temperature sensors, and humidity sensors embedded within composites provides valuable insights into how environmental factors affect both the structural and functional properties of the material [35].

3.6.2 Nanomaterial Enhancements

The incorporation of nanomaterials, such as CNTs, graphene, and nano-silica, into smart composites has been a significant advancement in enhancing the material's durability. Nanomaterials improve the mechanical properties, increase the thermal and UV stability, and offer self-healing capabilities, making them more resilient to environmental stresses [30].

3.7 Conclusion

Environmental and durability testing of smart functional composites is vital for ensuring their long-term reliability and performance in real-world applications. The review highlights the variety of testing methods used to evaluate the behavior of these materials under environmental stresses, including thermal cycling, humidity exposure, UV radiation, mechanical loading, and electromagnetic interference. Advancements in sensor technologies, nanomaterial integration, and self-healing capabilities have significantly improved the performance and longevity of these composites. However, continued research and field testing are essential for optimizing their durability and ensuring their widespread adoption in demanding applications.

References

1. Meng, Q. and Hu, J., A review of shape memory polymer composites and blends. *Compos. Part A: Appl. Sci. Manuf.*, 40, 11, 1661–1672, 2009.
2. Wang, Y., Pham, D.T., Ji, C., Self-healing composites: A review. *Cogent Eng.*, 2, 1, 2015, https://doi.org/10.1080/23311916.2015.1075686.
3. Khanbareh, H., Rasheed, A., Khaliq, J., 13 - Piezoelectric composites, in: *Organic Ferroelectric Materials and Applications*, Woodhead Publishing Series in Electronic and Optical Materials, pp. 457–475, 2022, https://doi.org/10.1016/B978-0-12-821551-7.00014-2.
4. Elhajjar, R., Law, C.T., Pegoretti, A., Magnetostrictive polymer composites: Recent advances in materials, structures and properties. *Prog. Mater Sci.*, 97, 204–229, 2018, https://doi.org/10.1016/j.pmatsci.2018.02.005.
5. Huo, B. and Guo, C.Y., Advances in Thermoelectric Composites Consisting of Conductive Polymers and Fillers with Different Architectures. *Molecules*, *27*, 20, 6932, 2022, https://doi.org/10.3390/molecules27206932.
6. Zhang, H., Dou, C., Pal, L., Hubbe, M.A., Review of electrically conductive composites and films containing cellulosic fibers or nanocellulose. *BioRes.*, 14, 3, 7494–7542, 2019.
7. Wang, Y., Wang, Y., Wei, Q., Zhang, J., Light-responsive shape memory polymer composites. *Eur. Polym. J.*, 173, 15, 111314, 2022, https://doi.org/10.1016/j.eurpolymj.2022.111314.
8. Santulli, C., 2 - Bio-inspired fiber composites, in: *Biomimetic Technologies, Principles and Applications*, pp. 33–51, Woodhead Publishing, Series in Electronic and Optical Materials, Elsevier, 2015, https://doi.org/10.1016/B978-0-08-100249-0.00002-1.
9. Narayana, K.J. and Burela, R.G., A review of recent research on multifunctional composite materials and structures with their applications. *Mater. Today Proc.*, 5, 2, 1, 5580–5590, 2018, https://doi.org/10.1016/j.matpr.2017.12.149.
10. https://www.precedenceresearch.com/composites-market
11. Ashrith, H.S., Jeevan, T.P., Xu, J., A Review on the Fabrication and Mechanical Characterization of Fibrous Composites for Engineering Applications. *J. Compos. Sci.*, 7, 252, 2023, https://doi.org/10.3390/jcs7060252.

12. Balachandar, M., Ramnath, B.V., Barath, R., Sankar, S.B., Mechanical Characterization of Natural Fiber Polymer Composites. *Mater. Today Proc.*, 16, 2, 1006–1012, 2019, https://doi.org/10.1016/j.matpr.2019.05.189.
13. Kolluru, R., Prasanthi, P., Kumari, A.S., Mechanical characterization of natural hybrid composites for automotive applications. *J. Braz. Soc. Mech. Sci. Eng.*, 45, 511, 2023, https://doi.org/10.1007/s40430-023-04442-0.
14. Olugbade, T.O. and Lu, J., Literature review on the mechanical properties of materials after surface mechanical attrition treatment (SMAT). *Nano Mater. Sci.*, 2, 1, 3–31, 2020, https://doi.org/10.1016/j.nanoms.2020.04.002.
15. Kumar, K.B., Anuradha, R., Dwivedi, S.P., Gupta, C., A.Al-Saady, F., Sethi, G., Design and Characterization of Multifunctional SMART Materials for Sensing and Actuation Applications. *E3S Web Conf.*, 430, 01132, 2023, https://doi.org/10.1051/e3sconf/202343001132.
16. Langat, R.K., De Luycker, E., Cantarel, A., Rakotondrabe, M., Toward the development of a new smart composite structure based on piezoelectric polymer and flax fiber materials: Manufacturing and experimental characterization. *Mech. Adv. Mater. Struct.*, 31, 27, 9345–9359, 2023, https://doi.org/10.1080/15376494.2023.2271746.
17. Kim, N.G., Han, M.W., Iakovleva, A., Park, H.B., Chu, W.S., Ahn, S.H., Hybrid composite actuator with shape retention capability for morphing flap of unmanned aerial vehicle (UAV). *Compos. Struct.*, 243, 112227, 2020, https://doi.org/10.1016/j.compstruct.2020.112227.
18. Chen, Q., Li, H., Zhang, Y., Thermal conductivity enhancement of graphene-based polymer composites for heat management applications. *J. Mater. Sci.*, 57, 6, 3417–3429, 2022.
19. He, J., Zhang, Z., Zhang, Y., Thermal and mechanical behaviors of shape memory polymer composites. *J. Smart Mater.*, 30, 4, 590–599, 2021.
20. Kim, J., Lee, S., Park, C., Thermal performance of self-healing composites under thermal stress: Infrared thermography study. *Compos. Sci. Technol.*, 199, 108265, 2021.
21. Li, Y., Wang, X., Liu, H., Thermal degradation and stability of self-healing composites. *J. Appl. Polym. Sci.*, 136, 35, 47801–47809, 2019.
22. Liu, H., Zhang, L., Zhang, W., Self-healing behavior of thermally activated composite materials. *Composites, Part A*, 133, 105841, 2020.
23. Wang, L., Zhang, Q., Li, T., Reversible thermochromic behavior of smart composites for energy-efficient windows. *J. Mater. Chem.*, 11, 13, 4035–4043, 2023.
24. Yang, Y., Liu, Y., Ma, Z., Thermal cycling effects on shape-memory composites. *Mater. Sci. Eng., A*, 715, 165–173, 2018.
25. Zhang, X., Li, B., Luo, Y., Thermal and mechanical performance of piezoelectric composites for energy harvesting. *J. Appl. Phys.*, 126, 6, 064501, 2019.
26. Zhang, Z., Zhang, Z., Wang, W., Thermal properties and shape recovery behavior of shape-memory polymer composites. *Polym. Test.*, 84, 106315, 2020.

27. Bhardwaj, R., Saha, S., Kaur, A., Thermal and mechanical stability of smart composites for aerospace applications. *J. Compos. Mater.*, 55, 4, 487–503, 2021.
28. Chakrabarti, A., Kumar, S., Ghosh, S., Self-healing composites: A review of damage detection and healing mechanisms. *Mater. Des.*, 211, 110051, 2022.
29. Mohan, A., Singh, M., Thakur, P., UV and moisture degradation of smart functional composites. *Polym. Degrad. Stab.*, 185, 109415, 2021.
30. Singh, S., Kamble, Z., Neje, G., Textile-reinforced composite structural-health monitoring systems. *Text. Prog.*, 56, 4, 415–516, 2024, https://doi.org/10.1080/00405167.2024.2422749.
31. Rahman, M., Saha, S., Patel, H., Mechanical and vibration testing of smart composites with integrated sensors. *J. Eng. Mater. Technol.*, 142, 3, 031010, 2020.
32. Shukla, D., Lee, K., Wang, Y., Accelerated aging of piezoelectric smart composites under environmental conditions. *J. Appl. Polym. Sci.*, 138, 17, 50051, 2021.
33. Saha, S., Singh, R., Kumar, P., Durability testing and environmental performance of smart composites. *Mater. Sci. Eng.*, 793, 139–150, 2020.
34. Wu, Y., Zhao, L., Zheng, X., Moisture-induced degradation and its effect on the electrical performance of CNT-based smart composites. *J. Nanosci. Nanotechnol.*, 19, 3, 1557–1563, 2019.
35. Zheng, H., Liu, Q., Wang, Y., Enhancement of UV resistance in smart composites using nanomaterials. *Polym. Compos.*, 42, 9, 4539–4549, 2021.
36. Zhao, Y., Li, J., Zhang, Y., Field testing and environmental performance of smart composites in wind turbine blades. *J. Renew. Sustain. Energy*, 14, 2, 023301, 2022.

4
Piezoelectric Nanocomposites

Param Singh*, Srijal Mishra, Roopak Varshney and Yajush Walia

Department of Mechanical Engineering, National Institute of Technology Hamirpur, Hamirpur, India

Abstract
Modern times are characterized by advanced materials and technology because of their unique benefits and qualities. Prior to now, only conventional piezoelectric materials have drawn the interest of the scientific community. However, due to their small size, high strength, and some other unique characteristics that piezoelectric nanocomposites exclusively possess, nanomaterials are now attracting attention on a global scale. Authors can create a piezoelectric compound by converting mechanical energy into electric energy according to specific mathematical equations, and as the generation changes, so do the products. This chapter covers the fundamentals of piezoelectric materials, their concepts, and a detailed discussion of piezoelectric nanocomposites, their properties, and the creation of a few nanocomposites. These materials have many applications, including manufacturing nano-transformers, portable nano-generators, harvesters, and other devices employed in the bio-medical field. Most piezoelectric nanocomposites are employed in the manufacturing of sensors. These applications have been discussed for a few of these types. As a result, it may be concluded that the technological world's future lies in compact objects with strong performance. This is also the direction that future research will go.

***Keywords*:** Nanocomposites, piezoelectricity, piezoelectric ceramics, piezoelectric polymers, piezoelectric transformer, energy harvesting

**Corresponding author*: psingh@nith.ac.in

Sandip Kunar, Pranav Charkha, Santosh Jaju and Harish Tiwari (eds.) Functional Composites: Role in Modern Engineering, (71–98)

4.1 Introduction

The word "piezo" comes from the Greek word "piezein" which means "to press." So, piezoelectricity, in short, means pressure electricity. Therefore, these are materials that, due to the piezoelectric effect, have the capacity to produce the electrical charges when mechanical pressure is applied. There is a complex definition of piezoelectricity, describing it as an electric polarization caused by mechanical strain like giving a certain kind of load or pressure onto some materials, so automatically the mechanical strength will be generated inside that material, so due to that the material is giving a certain kind of electricity, in some crystals belonging to a certain class of materials. Every material, if pressed, will not generate electricity; the materials should have that capability or properties to produce electricity that will also be discussed in this chapter.

The Curie brothers, Jacques and Pierre, investigated the influence of pressure on the creation of electrical charge by several natural crystals, such as quartz and tourmaline, in 1880. During this research, they found a phenomenon called piezoelectricity. So, basically, these are the materials that they have developed first and then they have seen these phenomena first time in the year 1880 and after that piezoelectric material came into the market. Now, scientists have done so much research on it to increase the efficiency of electricity. The evolution of piezoelectric composites has been marked by key milestones driven by visionary researchers. Beginning in 1971, Kitayama from NTT Japan pioneered the fusion of PZT powder and PVDF. The subsequent years saw Powell (1973) introducing PZT powder integrated with polyurethane, while Furukawa (1976) conceptualized a novel spherical model. Notably, Newnham (1978) at Penn State crafted coral replica composites from PZT and silicone rubber, emphasizing connectivity. In 1983, Banno's PT: polymer 0-3 type (NTK, Japan) broadened possibilities. Uchino's work in 1990 (Sophia University, later Penn State) led to carbon polymer piezo dampers and advanced PZT-based designs, including Terfenol magnetoelectric composites (2000) and energy harvesting composites (2004).

4.1.1 The Piezoelectric Effect

Piezoelectric crystals exhibit the piezoelectric action in two different ways.

4.1.1.1 Direct Piezoelectric Effect

The direct piezoelectric effect means that the mechanical stress applied to the crystal results in the generation of an electric charge (Figure 4.1).

4.1.1.2 Reverse Piezoelectric Effect

The reverse piezoelectric effect means a strain that is produced when an electrical field is applied to a crystal (Figure 4.2).

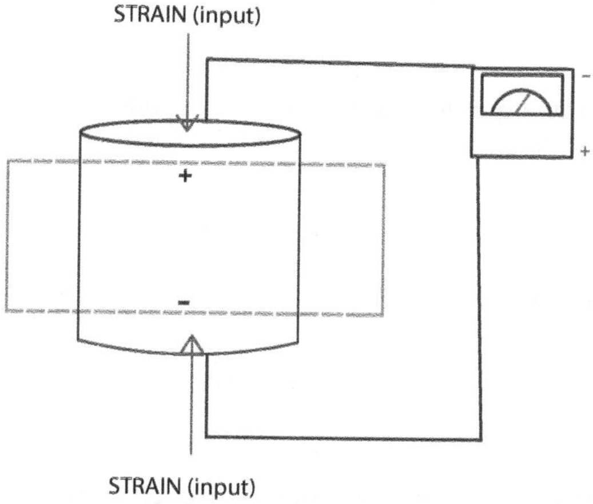

Figure 4.1 Schematic diagram of direct piezoelectric effect.

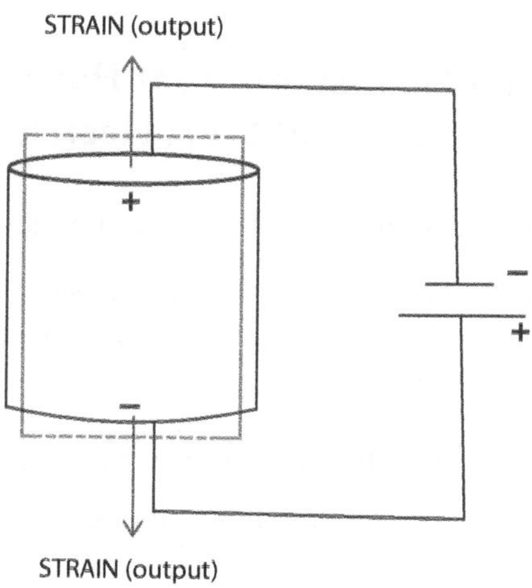

Figure 4.2 Schematic diagram of reverse piezoelectric effect.

These types of materials are really beautiful; as with applied pressure or a load, it will create electricity; at the same time, if electricity is applied to these materials, the materials will experience stress or possibly strain. As a result, it will alter in size and shape. Therefore, there will be some sort of internal deformation of the materials [1].

Therefore, the abovementioned two effects are essentially defined by the fundamental equations that connect:

E: the elastic variable field (E)
D: displacement to the elastic variable stress (T)
S: strain

So, the two-state equations are expressed as follows:
For the direct effect:

$$D = dT + \varepsilon^{T} ED$$

For the reverse effect:

$$S = s^{E} T + dE$$

where
d: charge constant as well as piezoelectric coefficient
s: elastic compliance
ε: permittivity/dielectric constant

4.2 Variables and Constants That Have an Impact on the Performance of Piezoelectric Materials

4.2.1 Electro-Mechanical Coupling Factors (k)

It serves as a gauge for how well a material transforms electrical energy into mechanical energy or mechanical energy into electrical energy.

For direct piezoelectric effect

$$k^2 = \frac{\text{Transformed mechanical energy into electrical energy}}{\text{input mechanical energy}}$$

which will give efficiency. So, how many input parameters are put inside the materials and as a result of the particular input parameters how much efficiency or may be output are getting, i.e., k^2 for direct piezoelectric effect [1].

For converse piezoelectric effect

$$k^2 = \frac{\text{Transformed electrical energy into mechanical energy}}{\text{input electrical energy}}$$

So, generally, k is denoted by k_{ij}, where I is electrical field direction and j is longitudinal vibration direction.

Depending on the structure, a typical piezoelectric ceramic can convert 25% to 50% of the energy provided to it in one form into another. For example, if applying pressure and that pressure is 100%, then 50% of that energy will be converted into electrical energy, resulting in a mechanical strain of 25% to 50%.

Therefore, that is dependent upon the crystal structure and material qualities. Therefore, high k is typically preferred for effective energy conversion.

4.2.2 Piezoelectric Strain (Charge) Constant (d)

It is measured in coulomb/newton (C/N) and represents the intercourse between the electrical charge produced per unit area and the applied force.

For direct piezoelectric effect

$$d = \frac{\text{the charge density (open circuit)}}{\text{applied stress}}$$

For reverse piezoelectric effect

$$d = \frac{\text{developing strain}}{\text{applied field}}$$

Where

$$d = k \text{ x } \sqrt{\varepsilon_o \varepsilon s^e} \text{ C/N}$$

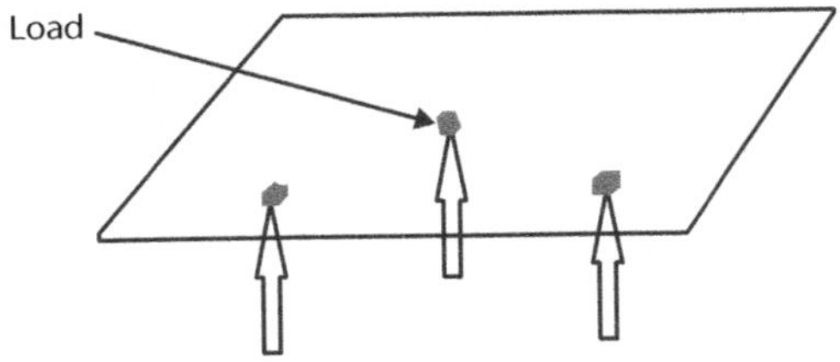

Figure 4.3 Load applied to the body.

k = electro-mechanical coupling factor
εo = relative permittivity of free space
ε = dielectric constant *(relative permittivity)*
se = compliance (10^{-12} m^2/N)

Generally, d is denoted by d_{ij} where i is (induced polarization direction) and j is stress applied direction. The d is a crucial sign of a material's viability for applications requiring strain-dependent actuators.

Essentially, what is happening is to check the material efficiency when applying pressure to various sections of an area, as shown in Figure 4.3, or whether applying load to a point or a whole area.

4.2.3 Piezoelectric Voltage Constants (g)

It is the reciprocal of the mechanical stress applied and the electrical field produced in Vm/N.

For direct piezoelectric effect

$$g = \frac{\text{field produced}}{\text{mechanical stress applied}}$$

For reverse piezoelectric effect

$$g = \frac{\text{strain produced}}{\text{Charge density applied}}$$

G_{ij} often stands in for g. "g" is a crucial constant for sensor applications.

4.2.4 Mechanical Quality Factor (Qm)

It is expressed as

$$Q_m = \frac{2\pi f\, Ue}{Pd}$$

Where
Ue: the system's accumulated mechanical energy
P_d: stands for power lost (watts)
The angular frequency is 2f.

U_e simply refers to the fact that a certain material has used up all mechanical energy prior to the application of any given load or pressure. Therefore, piezoelectric materials with high Qm are preferred because they are less likely to break under continual application of a little amount of load or pressure. Therefore, the material must possess a high degree of mechanical strength in order for it to constantly produce electricity when a load is applied.

4.2.5 Electrical Loss (tanδ)

The loss angle tangent is used to define the dielectric loss factor. The inverse of the gain factor is Q_m.

$$\text{Basically } \tan\delta = \frac{1}{Qm}$$

4.2.6 Dielectric Constant (ε)

The dielectric constant is defined as the proportion of a material's permittivity to that of free space.

Mathematically, the dielectric constant is

$$\varepsilon = \frac{Co\, h}{\varepsilon o\, A}$$

Where
ε: relative material permittivity
εo: relative permittivity, or 8.854 * 10-12 (F/m) of empty space
h: distance between electrodes (m^2)
Co: measured capacitance at 1 kHz (F)

4.3 Piezoelectric Nanocomposites

Piezoelectric nanocomposites are often made of ceramic and polymer materials, both of which have great promise due to their excellently customizable features. Because they are affordable and have desirable qualities, nanocomposites mostly based on polymers are employed extensively. Some advantages of this composite include high coupling factors, low acoustic impedance, good matching to water or human tissue, mechanical flexibility, a broad bandwidth combined with a low mechanical quality factor, and the capacity to make undiced arrays by simply patterning the electrodes. Three categories serve as the major divisions of nanocomposite materials. One is the 0-0 composite, which consists of two different types of nanoparticles with distinct components, phases, or types. The 0-2 composite is second in importance. The nanolamellar framework is generated from an experimental composite material created by uniformly

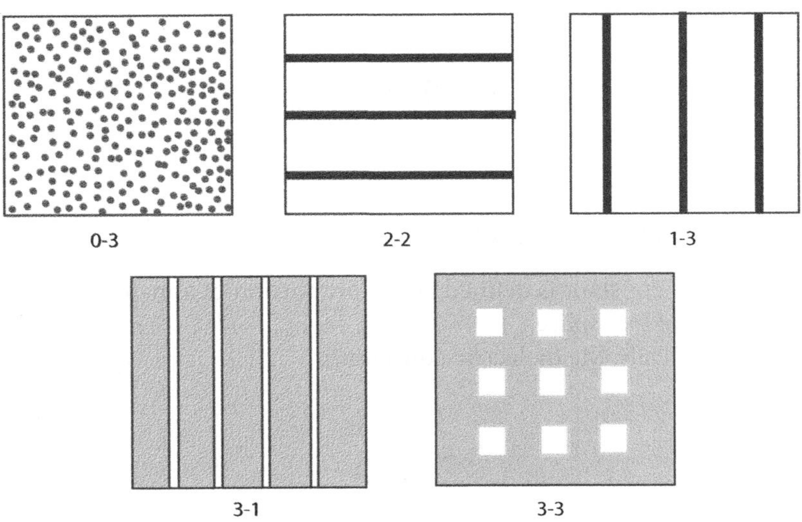

Figure 4.4 Ways of connection with their structure diagram.

or unevenly dispersing nanoparticles into two-dimensional elements. The nanoparticles are disseminated into a solid that is three-dimensional in the third process, known as 0-3 recombination [2, 3]. The composite piezoelectric materials consist a minimum of one piezoelectric material that has been blended with a non-piezoelectric material in accordance with a predetermined structural connection to create an advance material having a piezoelectric effect. In the case of composite materials, the ceramic phase and polymer phase are connected self in the shapes of 0 (particle), 1 (fiber), 2 (lamellar), and 3 (network), respectively. Figure 4.4 illustrates the typical connection mode [4].

Polymer-based composites are less denser than ceramics which is why they are widely used. Now let us discuss in detail what are nanocomposites based on polymer and ceramic [5, 6].

4.3.1 Piezoelectric Nanocomposite Materials—Polymer-Based

For sensor applications, PVDF and other polymer piezoelectric materials provide great choices. However, because of its extremely low elastic stiffness and low piezoelectric d constant, PVDF cannot be used alone to make actuators or high-power transducers. Transducers for applications like sonar that integrate sensor and actuator functions can be made with PZT polymer nanocomposites [7]. Although high piezoelectric ceramics also have some drawbacks because of their rigidity and weight, polymers are lightweight and flexible and therefore offer some advantages over ceramics in some applications. The electronic sector cannot function without them. For applications like sonar that require both actuator and sensor functions, PZT polymer composites are crucial in the construction of transducers [8].

Kitayama and Sugawara were the first who found the polymer composite 0-3 type (1972) [8].

Since then, research has been done on a variety of piezo composites, as summarized in Table 4.1. In Table 4.2, a number of composite piezoelectric materials are listed [9]. "d" is piezoelectric constant, which means the strain per unit electric field (in actuators application) which is one-tenth lower than the value of PZT. The piezoelectric "g" constant of PVDF, which represents the voltage per unit stress (in sensing applications) is 10 times greater than that of PZT, although this is because of the low dielectric constant. PZT shows a broad array of piezoelectric responses, but usually d is larger than PZT and g is smaller than PVDF. For underwater transducers that conduct simultaneously control and detection, based on a parameter, i.e., g^h, it is therefore shown that composite materials are much superior to single-phase materials like PZT or PVDF.

Table 4.1 On the basis of these values, materials are chosen for maximum efficiency [1, 46].

Piezoelectric constants of different materials						
Material parameters	**PZT - 4**	**$PbNbO_3$**	**PVDF**	**$LiNbO_3$**	**AIN thin film**	**PMN-PT**
Electro mechanical coupling (k)	0.47	0.33	0.19	0.16	0.24	0.57
Piezoelectric strain constant (d)	290	85	25	5.9	5.5	1,400
Piezoelectric voltage constant (g)	26	32	230	22	52	30
Mechanical quality factor (Qm)	High	Low	Low	Very high	Very high	Low
Electrical loss (tan δ)	0.004	0.01	0.3	0.001	0.0005	0.01
Relative permittivity at constant stress (ε)	1,270	300	8.4	29.8	12.0	3,950

Density: 10^3 kgm^{-3}, elastic constant in GPa and piezoelectric constants D_{33}: 10^{-12} CN^{-1}, g^{33} : 10^{-3} mVN^{-1} and gh : 10^{-3} mVN^{-1}.

Polymers are light and flexible and have certain advantages over ceramics, although high piezoelectric ceramics also have some disadvantages due to their rigidity and weight. They are crucial to the electronic industry. PZT composites are essential in the construction of transducers for devices like sonar, which serve as both actuators and sensors. Flexible piezoelectric materials have been the subject of current and continuing study, and polymer-ceramics systems are being researched. The piezo-responsiveness of the characteristics of polymer nanocomposites may change with the inclusion of tiny piezoelectric ceramic particles [10].

These materials have a bigger mechanical and piezo-response due to their size, which is substantially larger than that of micro-composites. The matrix of the polymer was given an additional chance to possess exceptional capabilities when nanoparticles were introduced. Because both the mechanical as well as electrical domains of these elements interact, piezoelectric materials have the capacity to transform mechanical energy or vibrations into electrical energy. For applications in devices, piezoelectric nanocomposites are employed in film and fiber form.

Table 4.2 Comparison of the piezoelectric response of PZT polymer composite with single phase materials, PVDF and PZT [9].

Material	Density (g/cm^3)	Elastic constant	Dielectric constant	D_{33} (10^{-3} C/N)	g^{33}	g^h (10^{-3} Vm/N)
PZT (501A) single phase	7.9	81	2,000	400	20	3
PZT : epoxy	3	19	400	300	75	40
PZT : silicone rubber	3.3	3	40	110	280	80
PZT : silicone rubber	4.5	19	400	250	60	-
PZT : PVDF	5.5	2.6	120	90	85	-
PZT : rubber	6.2	0.08	73	52	140	30
PZT : chloroprene rubber	-	-	40	-	-	90
Extended PVDF	1.8	3	13	20	160	80

Several ceramics have a higher piezoelectric coefficient compared to polymer compounds, which can minimize the frequency of battery replacement and electronic device use. However, the amount of piezoelectricity produced depends on how much the sample can be stretched. Ceramics are brittle, making it challenging to extend them appropriately. The most effective method is to concentrate on polymer–ceramic composites in order to produce a sample that has a better piezoelectric coefficient and good mechanical response, which might be utilized to make energy-collecting devices. The fact that the features of both ceramics and polymers in these materials are utilized is the main cause why scientists are focusing on ceramic–polymer piezoelectric nanocomposites rather than just ceramics or polymers [11].

4.3.2 Poling

By subjecting a material to elevated tempèratures along with a strong field of electricity, the poling process improves the characteristics of the piezoelectric material [12]. When a material's dipoles are randomly orientated, they cancel each other out to neutralize any noticeable piezoelectric action. Creating a field of electricity that is stronger than the saturation field but weaker than the breakdown field that causes the dipoles to align [13]. Despite being kept placed underneath the Curie temperature, the temperature is raised to help with the moving and rearranging of the dipoles. Dipoles within the material get aligned as a result, creating directionality. The material cools down while being exposed to a high electric field in order to maintain this effect. The two most widely used poling techniques are corona poling and contact poling as shown in Figures 4.5 and 4.6 [14].

A pair of electrodes are positioned between the solid material and heated *via* contact pooling. When an electric field is present, cooling takes place to

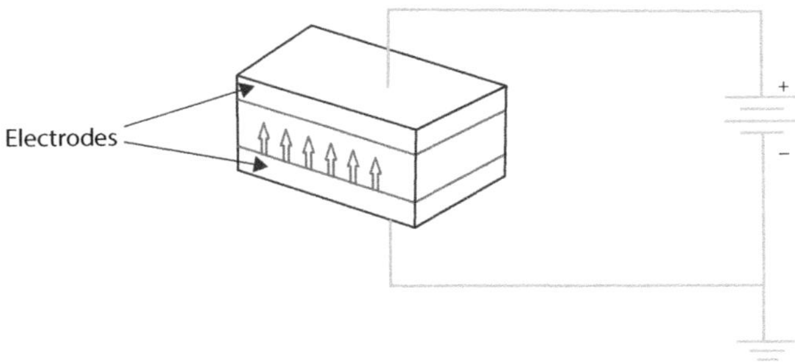

Figure 4.5 Schematic diagram of corona poling.

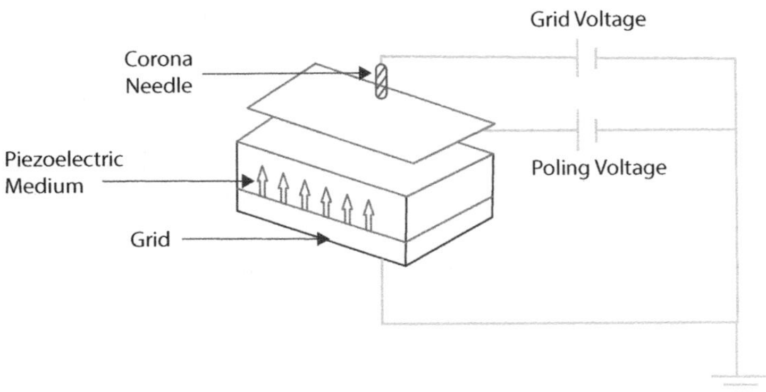

Figure 4.6 Schematic diagram of contact poling.

keep the dipole aligned [15]. While corona poling is being employed, only a single side of the material relates to an electrode and the other surface is free. The electrode supplies the material with a low voltage while keeping the pointy corona spike at an elevated voltage in the kilovolt zone atop the unbound surface. Gas molecules close to the extremely charged spike get ionized, rush in the direction towards the material's unbound surface, and then deposit there [16].

4.3.3 Preparation of a Nano-Polymeric Piezoelectric Composite

By incorporating nano yttrium oxide (Y_2O_3) at 3%, 6%, and 9wt% in the PVDF/PMMA (10/90) polymeric mixture, nano polymeric piezoelectric composites are made. The PMMA particles are then combined with various concentrations of nano Y_2O_3 using a magnetic stirrer at 80°C in a THF solvent, and the PVDF powder is combined for 1 h. As a result, a paste is created out of it, which is then molded to produce a thin film with the desired thickness. Then, several tests are carried out using a variety of equipment, such as the following: the wettability angle is measured using an optical dynamic or static contact angle meter and to investigate the morphology of produced films; a piezoelectricity test is carried out using a source meter, and optical microscopy and atomic force microscopy (AFM) are taken into consideration. PMMA, PVDF, and a mixture of 90% PMMA and 10% PVDF have hardness ratings of 71, 23, and 62, respectively. This indicates that the blend is immiscible and that the hardness of the combination does not conform to the mixture role. Adding Y_2O_3 causes a linear rise in hardness. This is a result of Y_2O_3's extreme hardness. Due to the polar nature of PVA and PVDF polymers, which can attract to the -OH group of water and reduce resistance as the relative humidity rises, the electrical resistance decreases as the relative humidity rises. Additionally, as humidity is absorbed by these polymers, the concentration of charge carriers rises, increasing conductivity. As the rate of adsorption rises, hydrogen ions also have a tendency to dissociate more quickly, which speeds up the release of electrons and improves conductivity. Particularly at ratios lower than 9% wt., the inclusion of Y_2O_3 had no impact on the PVDF/PMMA blend's structure or permeability. The losses factor for dielectric characteristics was inversely related to concentrations, and the dielectric constant abruptly decreased with frequency [17].

4.3.4 Piezoelectric Nanoparticle Polymer Composite Foam (PNPF)

A 1% wt. multiwall CNT PDMS and a 10% wt. BTO nanoparticle were mixed with a PDMS polymer to form foam structures. After a 5-h sonication session at room temperature, the composite solution is well dispersed. After being sonicated, the polymer resin and nanoparticles slightly changed into a viscous black liquid. Once a soft dough has formed, 1.8, 2.4, and 3.6 g of ultra-fine baker's sugar (with sugar particles no larger than 150 μm in diameter) are added to the resin in order to adjust the PNPF's porosity. Using a Teflon beaker and Kapton sheets, the soft dough is then rolled out to the necessary thickness of 350 μm. To guarantee complete curing, the sheets were then placed in an oven and the temperature was held at 100°C for 10 min. The cured sheets are then cleaned with ethanol and submerged in boiling water to remove the sugar powder before being dried in an oven set to 100°C. Using SEM and a methanol saturation method, the mechanical test was carried out to determine the elastic modulus and porosity of each foam, where the pore size is regulated by the sugar grain diameter [18, 19].

4.3.5 Reverse Effect of PNPF

PNPF device displacement is determined using a technique that mimics atomic force microscopy (AFM). When a laser is reflected on the back side of a cantilever beam, the positional shift of the beam that is reflected is monitored using a 1-D position sensitivity detection device (PSD). A voltage amplifier with a function generator is used to apply the external voltage once the sample has been attached to the measuring equipment. To show the piezoelectric phenomenon and not the static charge effect, a PNPF device and pure PDMS foam having 50% porosity are also examined.

It is required to separate piezoelectric and electrostatic actuation from the movement of the piezoelectric device because two conductive plates display electrostatic static actuation. It has been proved that the movement is not brought on by the foam framework itself when there is no measurable deformation in a pure PDMS device [20].

Three factors could account for this, namely:

- The less porous PNPF contains more active material than the more porous ones, the direct effects of less porous materials can be enhanced under the same mechanical stimulation due to lower stiffness, and the effect of void charge can be reduced with a decrease in the void fraction of the materials.

- The void charge piezoelectric effect, which is a pseudo piezoelectric effect brought on by the deposited charge in the pores, can be decreased whenever the variation in the dielectric coefficient in the composite elements and the void space is lowered and the fraction of the void space is raised [21].
- The surface charge of the active material could cause the empty space charge that PNPF could produce; however more research is needed to completely grasp the PNPF's electromechanical operation, transduction effectiveness, and frequency responsiveness.

4.4 Piezoelectric Polymer Materials

There are various polymer materials used in piezoelectric materials fabrication; the most commonly used are discussed below.

4.4.1 Polyvinylidene Fluoride (PVDF)

The PVDF compound was first identified in 1969. It is a type of semi-crystalline polymer, and its crystallites are made up of chains that are oriented to generate dipoles. Then, an amorphous matrix surrounds this crystalline structure. PVDF is a strong piezoelectric polymer; however, certain requirements must be satisfied for this polymer to show adequate piezoelectricity.

These requirements are as follows:

- The resultant dipole force of a chain of polymers should be substantial and parallel to the chain axis.
- Polymers must form polar crystals, and the polar axis of these crystals must stay parallel to the chain axis of the polymer.
- For samples made of film or fiber, the molecular chain must be aligned perpendicular to the film or fiber dimension.
- It is ideal for the polymer to have an elevated crystallinity, and the polar axis will be oriented perpendicular to the direction in which the sample's thickness is measured [22].

4.4.2 Polyvinylidene Fluoride Trifluoro Ethylene (PVDF-TrFE)

A variety of applications, such as memory that is nonvolatile in organic electronic systems, of PVDF and PVDF-TrFE have been thoroughly explored. Their properties include ferroelectricity, piezoelectricity, and pyroelectricity. PVDF lacks the crystallinity that the copolymer TrFE demonstrates. PVDF lacks the crystallinity that the copolymer TrFE demonstrates. By heating this copolymer to a temperature between Curie's transition temperatures (Tm), it is simple to convert it into the beta phase using heat treatment [23].

4.4.3 Polyvinylidene Cyanide-Vinyl Acetate

In a copolymer of vinyl cyanide (VDCN) and vinyl acetate with the chemical formula $(CH_2\text{-}C(CN)_2\text{-}H_2\text{-}CHOCOCH_3)n$, these polymers have cyano groups, a significant dipole moment, and considerable piezoelectricity. These types of polymers are amorphous, and their piezoelectricity differs from that of crystalline polymers. The arrangement of the molecules individually governs piezoelectricity in crystalline polymers. It will have a significant dipole moment and thus good piezoelectric properties if the chain of polymers is following the path of the molecular rotational movement [24].

4.4.4 Polyamide 11 (Nylon 11)

This polymer's primary quality is excellent thermal stability; in addition, it has outstanding mechanical, ferroelectric, and piezoelectric capabilities, and it has very less water absorption. Compared to other piezoelectric polymers, Nylon 11 is more thermally resistant [25].

4.4.5 Cellular Polypropylene (PP)

Ferro-electrets with cellular polypropylene (PP) are some of the most efficient materials. In Kirjavainen's disclaimer, it is stated that this polymer has an unusually high d_{33} value when assessed approaching up to 1,200 pC N-1 for an individual film and a range between 130 and 2,100 pC N-1 for several layers. In addition to having a small modulus of rigidity of about 1 MPa, cellular PP also offers good fatigue resistance and is inexpensive [26]. Additionally, it has been demonstrated that the pressure exerted directly proportionally affects the piezoelectric response of cellular PP. These components have been used in accelerometers, transducers, and microphones.

4.4.6 Poly-Organo-Phosphazenes (POPh)

The piezoelectric characteristics/properties of poly-organo-phosphazenes (POPh), which are straight organo-polymers, are due to POPh's mesophase. POPh's piezoelectric characteristics are improved by fluorination because dipoles generated by electronegative fluorine atoms can align in response to the application of mechanical force. Solvent casting on a glass support under argon is one technique for fabricating POPh thin films, followed by vacuum drying. In order to get the chains to align more favorably, annealing can be used to increase the piezoelectricity of POPh, or it can be doped with other ions. The structure's piezoelectric capabilities can be considerably enhanced by adding $LiSO_3CF_3$ salt [27].

4.5 Piezoelectric Nanocomposite Materials—Ceramic-Based

Due to their availability, stability, and superior thermal performance, ceramics are commonly employed as a class of piezoelectric materials. Most ceramics have a perovskite structure [28]. Lead-containing compounds are frequently found in piezoelectric ceramics, which mostly have a perovskite structure. However, lead-free piezoelectric ceramics have been created to minimize the dangerous features of lead compounds, such BT ($BaTiO_3$) and potassium sodium niobate (KNN/$K_{0.5}Na_{0.5}NbO_3$). These materials provide a number of benefits, including superior piezoelectric capabilities, inexpensive manufacturing costs, and straightforward fabrication processes and are able to transform themselves into a multitude of forms. Because of this, ceramics are commonly used in a variety of applications, such as transducers, sensors, and actuators [29]. Some of the common piezoelectric ceramics are given below.

4.5.1 Lead Zirconate Titanate (PZT)

Pb (Zr1x Tix)O_3, or PZT, is a perovskite-structured ceramic material with outstanding properties. In the early 1950s, it was originally found at the Tokyo Institute of Technology. Due to PZT's strong piezoelectric constant (360 pC N1) and electromechanical coupling coefficient (0.69), more mechanical energy is converted to electrical energy. Additionally, it has a Curie temperature of 350°C and the unusual feature of maintaining its piezoelectric properties at high temperatures [30]. Since PZT is also physically robust, chemically stable, and does not degrade in a damp environment, it may be used in a range of applications. With applications ranging from atomic force microscopy to aircraft vibration management, PZT has grown to be the most commonly used piezoelectric material. PZT's performance is its greatest advantage; average voltage outputs can reach 100 V, albeit current output is on the range of nano and milli-amps. PZT is not a good fit for applications that need flexibility, which is one of its main drawbacks. It is brittle and unable to endure significant deformation due to its good density (65 GPa) and high Young's modulus [31].

4.5.2 Potassium Sodium Niobate (KNN)

The first perovskite-structured ferroelectric substance, known as $BaTiO_3$ or BT, was found by scientists during World War II and was widely employed in phonograph needles and sonar detection. It gained notoriety at first for having a large capacitance (a dielectric constant of 1,100) [32]. In comparison to PZT, pure, unaltered BT exhibits a low piezoelectric constant (191 pC N1), a low voltage constant (11.4103 V m N1), and a low Curie temperature (130 C), but BT ceramics' electromechanical qualities have been substantially improved, resulting in piezoelectric capabilities that are on par with PZT. The piezoelectric characteristics of BT have dramatically improved as a result of domain size reduction [33].

4.5.3 Bismuth Sodium Titanate (BNT)

$Bi_{0.5}Na_{0.5}TiO_3$ is often a lead-free piezoelectric ceramic. BNT is a ceramic without lead that Smolensky and colleagues discovered in 1960 and has exceptional piezoelectric characteristics. Due to its strong strain response, it has been shown to be beneficial in actuator applications. It has been demonstrated that BNT's piezoelectric characteristics can be improved by adding dopants [34].

4.5.4 Aluminum Nitride (AlN)

Another lead-free alternative to PZT is aluminum nitride, which has a wurzite crystal structure. Because of its advantages—improved complementary metal oxide semi-conductor manufacturing compatibility, resilience withholding humidity, and a Curie temperature of more than 1,000°C—it is primarily used in microelectromechanical system (MEMS) devices [35].

4.5.5 Lithium Niobate

The substance lithium niobate ($LiNbO_3$) is also a lead-free ceramic with a Curie temperature of 1,000°C. High temperature transducers frequently employ this material because of its robustness at high temperatures. It has become well-liked as a material for MEMS resonators because of its exceptional electromechanical coupling and exceptionally high quality-factor [36].

4.6 Improvements to Piezoelectric Ceramics

It was initially investigated in 1954 for use in capacitors, (Ba, Ca)(Zr, Ti)O_3, or BCZT. However, it was not until 2009 that researchers discovered that BCZT had exceptional piezoelectric characteristics, such as an extremely high piezoelectric coefficient on the same level with or even higher than PZT (the value varies with stoichiometry and grain size). As a result, BCZT is now a leading lead-free piezoelectric ceramic material on the market. It has been demonstrated that dopants enhance BCZT characteristics. When BCZT and BaTi0.89Sn0.11O_3 (BTSn) were coupled, Merselmiz *et al.* demonstrated that the properties were improved and that the addition of 0.2BTSn to the system resulted in a d_{33} value of 228 pC N1 [37].

4.7 Applications of Piezoelectric Nanocomposites

4.7.1 Bio-Medical

Due to their ability to provide the necessary electrical output when mechanically stretched and their potential to stimulate axonal outgrowth and brain tissue regeneration, piezoelectric composites are now improving neural cell function activities [38].

4.7.2 Piezoelectric Tactile Sensors

These sensors that monitor force or pressure have been made using piezoelectric polymers and nanocomposites made of polymers. Due to its mechanical strength, lightweight, low power consumption, biocompatibility, and other benefits, PVDF is a possibility for use in these applications. It has both piezoelectric and pyroelectric capabilities. Additionally, PVDF has high dielectric permittivity, which increases polarization and dipole moment and, consequently, piezo and pyro characteristics [7]. The permittivity, mechanical toughness, thermal stability, piezoelectricity, and other properties may be improved by combining nanoparticles such as zinc oxide or $BaTiO_3$ nanoparticles. Therefore, the manufacturing of piezoelectric/pyroelectric sensors requires the use of additional PVDF-based piezoelectric polymer nanocomposites.

4.7.3 Piezoelectric Vibrational Energy Harvestors Using Polymer Nanocomposite

Today, a variety of piezoelectric materials are accessible for use in power harvesting applications. It is possible to modify materials by, for example, doping the host in the case of piezo-ceramics, integrating nano-sized ceramics in a ferroelectric polymer, altering the electrode pattern, changing the poling condition and stress direction, applying prestress to increase coupling and applied strain, tuning the resonant frequency of the device, etc. [39]. Currently, researchers are focusing on increasing the efficiency of power harvesting using these piezo-electronic materials [40]. $BaTiO_3$, PZT, $LiNbO_3$, $CaTiO_3$, ZnO, and other materials are examples of piezoelectric ceramics. PVDF, odd Nylons (such as nylon 11, nylon 9, and others), and PVC are examples of piezoelectric polymers.

In comparison to pristine PVDF, nylons, PVCs, and other materials, PVDF copolymers exhibit improved piezoelectric properties. PVDF-TrFE is the best performing PVDF copolymer in this regard. In order to meet the needs of numerous applications, researchers are working to produce piezoelectric films, fibers, and solid samples [41]. Sensing, energy harvesting, and actuation are just a few of the numerous uses for piezoelectric polymers. Smart clothes, watch straps, key straps, necklaces, belts, and shoes are just a few of the applications that may be used, some of which are now in development. However, piezoelectric polymers must overcome a few obstacles for integration to be successful, namely:

- Torsion and rough handling
- Piezoelectric characteristics degrade as temperature rises
- Washing (clothing)

4.7.4 Piezoelectric Nano-Generator (PENG)

Figure 4.7 illustrates a piezoelectric nano-generator, an energy harvesting device which has the capacity to convert environmental mechanical energy

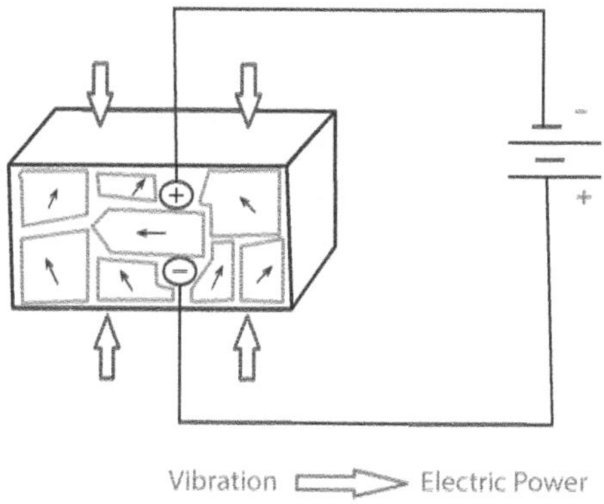

Figure 4.7 Schematic diagram of a piezoelectric nano-generator.

into electrical energy. Accordingly, PENG can be classified into the following categories:

I. Force applied perpendicular to the nanowire's axis
II. Force applied parallel to the nanowire's axis

In this specific instance, pressure or perhaps vibrations are applied to the materials; as a result, electricity is being produced. Now, in that specific situation, the stress and strain directions are determined by how the materials are stored, whether storing them horizontally or vertically.

4.7.4.1 Force Applied Perpendicular to the Nanowire's Axis

As a result, the entire piezoelectric structure deforms when an outside force is used to it by moving point in perpendicular direction. Therefore, the stretch region with the positive strain will show positive electrical potential. As tilt to a particular material, tension acts in these sides while compression acts in these sides, so the confined section with the strain negatively will display the negative electrical potentials.

As depicted in Figure 4.8, one will thus extend and the other will compress. The Ohmic contact that forms between the nano-structures and the bottom metal electrode balances the electric field generated at the nanostructure's tip.

The Ohmic effect is known as this. So, this is where it truly works. The formation of the Schottky contact, also known as the Schottky behavior, between the top metal electrode and the nanostructure's tip is what causes the development of the electric current.

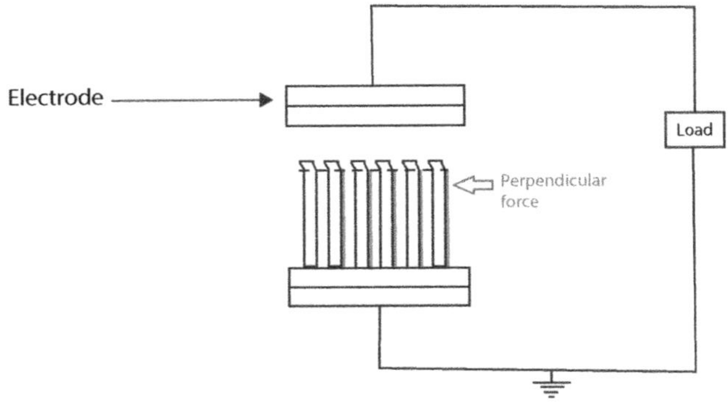

Figure 4.8 Force acting perpendicular to the nanowire's axis.

4.7.4.2 *Force Applied Parallel to the Nanowire's Axis*

As a result, in this specific instance, when the force is directed toward the nanowire's tip, uniaxial compressive is produced in the nanowires because there is pushing of the materials in the same way. The Schottky obstacle or possibly the Schottky contact is known in this particular case as the Ohmic contact. As a result of the particular use of force, the material between the two electrodes is compressed, as shown in Figure 4.9. Therefore, the nanowire's tip is going to have a negative piezoelectric potential as a result of the piezoelectric action. As a result of electrons moving from the top to the bottom, a positive electric potential is generated at the other tip.

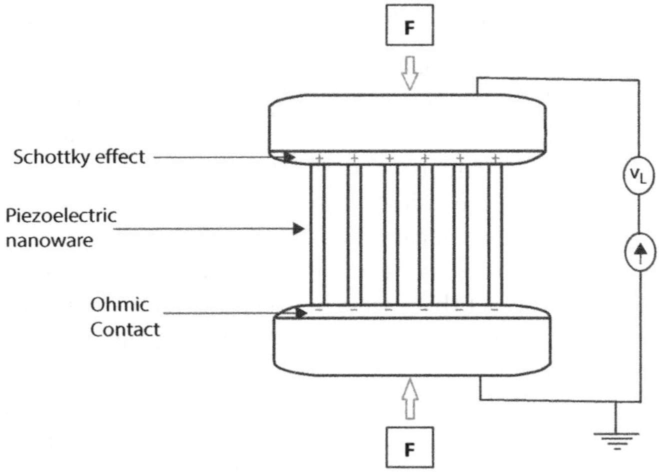

Figure 4.9 Force applied parallel to the nanowires' axis.

Therefore, the piezoelectric effect produced immediately disappears when the exterior force is withdrawn.

As a result, the pressure is relieved, and there will no longer be any generation of electricity. The movement of collected electrons is across the circuit exterior from the nanostructure's bottom electrode to the top electrode, and therefore through this circuit, neutralizes the positive potentials at the nanostructure's lowermost point. Thus, they are counteracting one another [42].

4.7.5 Nanocomposite Electrical Generators (NEG)

The electromechanical capabilities of the material and the Schottky contact that forms between the nanostructure's tip and the metal electrode are the key driving forces behind the power production process. Due to their brittleness and poor formability, inorganic materials are only useful in a few applications [43]. The necessity for materials with ductility arises as a result. Therefore, polymers, particularly PVDF, are employed as alternatives to metals in applications where an efficient generator design is necessary.

Piezoelectric materials have three major types based on the unit cell structure: the perovskite class (ABO_3), the bismuth layer-structure class ($Bi_2Ax\text{-}1BxO_3x\text{+}3$), and the pyrochlore class ($A2B_2O_7$) [44]. NEG is a three-dimensional configuration made up of these three components. The authors used two metal plate electrodes in this situation, as indicated in Figure 4.10. Now, among these, it resembles composites. Both the polymer matrix and the nano-structured components are present in composites. Next, the polymer matrix and piezoelectric nanostructure were produced vertically.

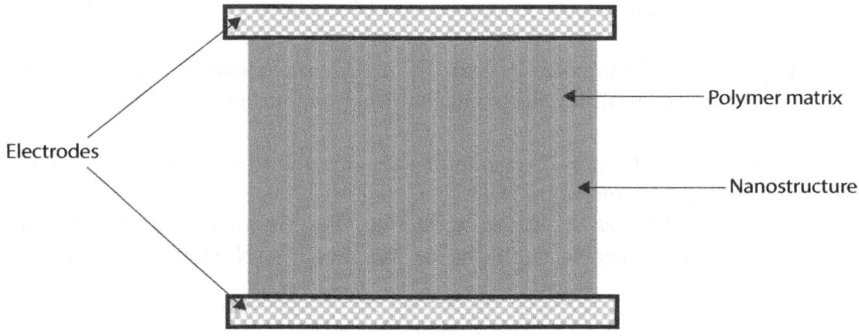

Figure 4.10 Schematic diagram of nanocomposite electrical generators.

The voids within piezoelectric nanostructures are filled with polymer matrix. Thus, this resembles a pillar, and the polymeric matrix fills the space between the pillars. An AFM tip can bend a zinc oxide nanowire in a NEG, which has a greater efficiency than the original nano-generator designs and also offers a more sustainable energy source [45].

References

1. Covaci, C. and Gontean, A., Piezoelectric Energy Harvesting Solutions: A Review. *Sensors*, 20, 12, 3512, 2020, https://doi.org/10.3390/s20123512.
2. Kong, X.L., Liu, Y.B., B.Y., Research progress in nano-composite materials. *Mater. Sci. Technol.*, 10, 4, 436–441, 2002.
3. Pan, J.S., Huang, X.H., Gu, S.X., Classification of Nanometer Materials and Its Basis. *Mater. Rev.*, 14, 11, 28–29, 2000.
4. Li., C., *Preparation and properties of micro-nano piezoelectric composites*, University of Electronic Science and Technology of China, China, 2016.
5. Lee, M., Chen, C.Y., Wang, S., Cha, S.N., Park, Y.J., Kim, J.M., Chou, L.J., Wang, Z.L., A Hybrid Piezoelectric Structure for Wearable Nanogenerators. *Adv. Mater.*, 24, 13, 1759–1764, 2012.
6. Anton, S.R. and Sodano, H.A., A Review of Power Harvesting Using Piezoelectric Materials (2003–2006). *Smart Mater. Struct.*, 16, 3, R1, 2007.
7. Sirohi, J. and Chopra, I., Fundamental Understanding of Piezoelectric Strain Sensors. *J. Intell. Mater. Syst. Struct.*, 11, 4, 246–257, 2000.
8. Kitayama, T. and Sugawara, S., Study Comm. Electronic Circuit Components & Mater, Ref. No. CPM72-17, 1972.
9. Abe, K., Uchino, K., Nomura, S., The Electrostrictive Unimorph of Dispacement Control, *Jpn. J. Appl. Phys., 21, L408*, 2, 81, 1982.
10. Chen, H., Zheng, Z.Q., Qi, Y.J., Preparation and electrical properties of flexible PZT epitaxial thin films. *J. Hubei Univ.*, 41, 3, 277–281, 2019.
11. Manbachi, A. and Cobbold, C., Development and Application of Piezoelectric Materials for Ultrasound Generation and Detection. *Ultrasound*, 19, 4, 187–196, 2011.
12. Ramadan, K.S., Sameoto, D., Evoy, S., A review of piezoelectric polymers as functional materials for electromechanical transducers. *Smart Mater. Struct.*, 2014.
13. Tressler, J.F., Alkoy, S., Newnham, R.E., Piezoelectric sensors and sensor materials. *J. Electroceram.*, 1998.
14. Dargaville, T.R., *et al.*, Characterization, Performance and Optimization of PVDF as a Piezoelectric Film for Advanced, SAND2005-6846 (Sandia), 2005.

15. Hoshyarmanesh, H. and Maddahi, Y., Poling process of composite piezoelectric sensors for structural health monitoring: a pilot comparative study. *IEEE Sens. Lett.*, 2018.
16. Park, C., Ounaies, Z., Wise, K.E., Harrison, J.S., *In situ* poling of amorphous piezoelectric polyimides. *Polymer*, 2004.
17. Fabrication and characterisation of piezoelectric polymer Nano Composites for Humidity Sensors Applications. *Int. J. Eng. Technol.*, 8, 1.5, 434–439, 2019.
18. Peng, S.H., Harley, P.G., Huges, T.C., Guo, Q.P., Controlling morphology and porosity of porous siloxane membranes through water content of precursor micromemulsion. *Soft Matter*, 2012.
19. Fan, F.R., Lin, L., Zhu, G., Wu, W.Z., Zhang, R., Wang, Z.L., Transparent Tribo-electric Nano-generators and Self-Powered Pressure Sensors Based on Micro-patterned Plastic Films. *Nano Lett.*, 2012.
20. Pister, K.S.J., Fearing, R.S., Howe, R.T., A Planar Air Levitated Electrostatic Actuator System. *Micro Electro Mechanical Systems*, pp. 67–71, 1990.
21. Arvelo, J.I., Optimizing Voided Piezoelectric Polymers For Acoustic Sensors. *AIP Conf. Proc.*, 1140, 140–155, 2009.
22. Kawai, H., The piezoelectricity of polyvinylidene fluoride. *Jpn. J. Appl. Phys.*, 1969.
23. Harrison, J. and Ounaies, Z., *Piezoelectric polymers*, pp. 23681–2199, ICASE, NASA Langley Research Center Hampton, Virginia, 2001.
24. Persano, L., Dagdeviren, C., Su, Y., Zhang, Y., Girardo, S., Pisignano, D., Huang, Y., Rogers, J.A., High performance piezoelectric devices based on aligned arrays of nanofibers of poly(vinylidenefluoride-co-trifluoroethylene). *Nat. Commun.*, 2013.
25. Simpson, J., Ounaies, Z., Fay, C., Polarization and piezoelectric properties of a nitrile substituted polyimide. *Mater. Res. Soc. Symp.*, VA 23681, 6, 1997.
26. K. Kirjavainen, Electromechanical Film and Procedure for Manufacturing Same. US Patent US4654546, 1984, 1987.
27. Kulichikhin, V., Semakov, A., Frenkin, E., Shabeko, A., Tur, D., Linear polyorganophosphazene films as flexible piezoelectrics and actuators. *Polym.*, 2015.
28. Jaffe, B., *Piezoelectric Ceramics*, vol. 3, Elsevier, Amsterdam, 2012.
29. Hooker, M.W., Properties of PZT-Based Piezoelectric Ceramics Between –150 and 250, NASA/CR-1998-208708, NAS 1.26:208708, 1998.
30. Low, T.S. and Guo, W., Modeling of a three-layer piezoelectric bimorph beam with hysteresis. *J. Microelectromech. Syst.*, 1995.
31. Smeltere, I., Antonova, M., Dunce, M., Livinsh, M., Garbarz-Glos, B., Electrical and mechanical properties of KNN based lead-free ceramics. *12th Conf. of the European Ceramic Society – ECerS*, XII vol. 1, pp. 5–7, 2011.

32. Li, P., Huan, Y., Yang, W., Zhu, F., Li, X., Zhang, X., Shen, B., Zhai, J., High-performance potassium-sodium niobate lead-free piezoelectric ceramics based on polymorphic phase boundary and crystallographic texture. *Acta Mater.*, 2019.
33. Lisińska-Czekaj, A., Czekaj, D., Garbarz-Glos, B., Bąk, W., Zate, T.T., Jeon, J.-H., Dielectric Spectroscopy Studies and Modelling of Piezoelectric Properties of Multiferroic Ceramics. *Appl. Sci.*, *13*, 7193, 2023.
34. Reichmann, K., Feteira, A., Li, M., Bismuth sodium titanate based materials for piezoelectric actuators. *Materials*, 2015.
35. Gerfers, F., *et al.*, Sputtered AlN thin films for piezoelectric MEMS devices - FBAR resonators and accelerometers, in: *Solid State Circuits Technologies*, 2010.
36. Wang, R., Bhave, S.A., Bhattacharjee, K., Design and fabrication of S0 Lamb-wave thin-film lithium niobate micromechanical resonators. *J. Microelectromech. Syst.*, 2015.
37. Guan, S., Yang, H., Cheng, S., Tan, H., Qiao, G., Chen, Q., Xu, J., Yuan, L., Wang, X., Yang, L., High-Temperature Piezoelectric Response and Thermal Stability of BiGaO3 Modified BiFeO3–BaTiO3 Lead-Free Piezoelectric Ceramics. *Crystals*, *13*, 1026, 2023.
38. Wang, S.D., Chen, C.L., Wang, J., Li, C.-B.-W., Zhou, J.L., Liu, Y.-X., Jiang, Y.-Q., Zhu, L.F., Li, C., Gong, W., Guo, W., Tang, X.D., Yao, F.-Z., Wang, K., Synergetic chemo-piezodynamic therapy of osteosarcoma enabled by defect-driven lead-free piezoelectrics. *Adv. Funct. Mater.*, 32, 44, 2208128, 2022.
39. Wu, J., Fan, Z., Xiao, D., Zhu, J., Wang, J., Multiferroic bismuth ferrite-based materials for multifunctional applications: Ceramic bulks, thin films and nanostructures. *Prog. Mater Sci.*, *84*, 335–402, 2016.
40. Wu, N., Bao, B., Wang, Q., Review on engineering structural designs for efficient piezoelectric energy harvesting to obtain high power output. *Eng. Struct.*, *235*, 112068, 2021.
41. Chen, Z., Xia, Y., He, J., Xiong, Y., Wang, G., Elastic-electro-mechanical modeling and analysis of piezoelectric metamaterial plate with a self-powered synchronized charge extraction circuit for vibration energy harvesting. *Mech. Syst. Signal Process.*, *143*, 106824, 2020.
42. Zeng, S., Zhang, M., Jiang, L., Wang, Z., Gu, H., Xiong, J., Du, Y., Ren, L., Wearable Piezoelectric Nanogenerators Based on Core–Shell Ga-PZT@GaOx Nanorod-Enabled P(VDF-TrFE) Composites. *ACS Appl. Mater. Interfaces*, 14, 6, 7990–8000, 2022.
43. Veeralingam, S., Nanda, O.P., Badhulika, S., Lead-free Bi2CuO4 interspersed into PDMS matrix-based bifunctional piezoelectric nanogenerator for vibrational energy harvesting and visible light photodetection applications. *J. Alloys Compd.*, 961, 171127, 2023.
44. Pawar, O.Y., Patil, S.L., Redekar, R.S., Patil, S.B., Lim, S., Tarwal, N.L., Strategic Development of Piezoelectric Nanogenerator and Biomedical Applications. *Appl. Sci.*, 2023.

45. Lu, Y., Jiang, L., Yu, Y., Wang, D., Sun, W., Liu, Y., Yu, J., Zhang, J., Wang, K., Hu, H., Wang, X., Ma, Q., Wang, X., Liquid-liquid triboelectric nanogenerator based on the immiscible interface of an aqueous two-phase system. *Nat. Commun.*, 2022.
46. Fraga, M., Furlan, H., Pessoa, R., Massi, M., Wide bandgap semiconductor thin films for piezoelectric and piezoresistive MEMS sensors applied at high temperatures: An overview. *Microsyst. Technol.*, 20, 9–21, 2013, 10.1007/s00542-013-2029-z.

5

Modulation of Waveform Effect on Ni/Nano-ZrO_2–TiO_2 Composite Coating on Mild Steel

Koona Bhavani[1], VSN Venkata Ramana[1], Chitrada Prasad[2*], A. Ramesh[3] and Marana Lalitha[4]

[1]*Department of Mechanical Engineering, GITAM Institute of Technology, GITAM (Deemed to be University), Visakhapatnam, Andhra Pradesh, India*
[2]*Department of Physics, Aditya University, Surampalem, Andhra Pradesh, India*
[3]*Department of Electrical and Electronics Engineering, Aditya University, Surampalem, Andhra Pradesh, India*
[4]*Department of Chemistry, Aditya University, Surampalem, Andhra Pradesh, India*

Abstract

In this work, electrodeposition has been carried out galvanostatically by the periodic modulation of rectangular and triangular waveform, and the waveforms' effect in the process of conventional electroplating of nickel/ZrO_2–TiO_2 nanocomposites was studied. Deposition conditions, including pulsing power density (pd) and layering degree, are optimized to enhance coating performance against corrosion by electrodynamic polarization test at 3.5 wt.% NaCl environment. Analysis of corrosion behavior across different pulse shapes considers surface morphology and phase structures, utilizing SEM and XRD techniques. ZrO_2–TiO_2 nanocomposite additives and waveform effect in deposition produce improved corrosion protection efficiency and structures of smooth morphology. The observations highlight a finer grain size in deposits obtained *via* triangular pulsing current and also influenced by nano ZrO_2–TiO_2 particles affecting the interface creation between Ni coating and substrate and grain growth energetics. Overall, coatings deposited at the same peak current density using triangular pulses demonstrate superior corrosion resistance compared to those from rectangular pulses.

**Corresponding author*: chitrada.prasad@gmail.com

Sandip Kunar, Pranav Charkha, Santosh Jaju and Harish Tiwari (eds.) Functional Composites: Role in Modern Engineering, (99–114) © 2025 Scrivener Publishing LLC

Keywords: ZrO_2–TiO_2, nanocomposites, rectangular, triangular, waveform, pitting corrosion

5.1 Introduction

Mild steel is broadly used in the fabrication of various metal equipment due to easy cast ability, formability, and weldability. Even though mild steel is processed with easy fabric ability and lower cost, it suffers from poor corrosion resistance and lower surface strength [1–3]. To improve the corrosion resistance and surface protection ability of steels [4], coatings are more reliable in recent days [5, 6]. Pulsed electrodeposition [7] process has more advantages over the continuous process, handles more reliability, and has normal working pressure, homogeneous distribution, low-cost maintenance, and reduced wastage compared with other coating methods like sol-gel dip coating [8], physical vapor deposition (PVD) [9], spraying [10], etc. In this process of pulse current deposition, higher rate of grain nucleation and more refined grain structures were obtained in coatings [11]. Previous research was focused on fabricating surface coatings with pulsed electrodeposition process with the direct, rectangular [2], and reverse pulse current [12] having been used in the process of nickel, zinc, copper, and its alloys' deposition. In general, during recent years, nickel plating is broadly used in automobile and defense applications to enhance the structural strength and tribological efficiency of engine parts [13–15]. The nickel-coated steel materials exhibit highest corrosion resistance in aggressive environments and are also used as radar absorbing materials [16]. The electro-co-deposition of second-phase nanoparticles in the nickel electrolyte bath such as TiO_2 and ZrO_2 is very effective to produce metal matrix composite (MMC) coatings through ionic diffusion of metallic and non-metallic particles with pure metals or alloys. Therefore, the mechanical, tribological, and corrosion resistance properties can be improved. In a pulse electrodeposition process, ions are migrated toward the electric field by the charge diffusion that takes place and incorporation of crystal ions of the cathode material on anode material during surface diffusion [17].

While extensive research has been carried out and certain fundamental principles of electrical co-deposition have been established, there remains a gap in the understanding of the deposition behavior and morphology of Ni when combined with nano-sized ZrO_2–TiO_2 electro-composites.

This paper aims to study the morphological characteristics of the composites produced under both rectangular and triangular waveforms, including relaxation time, utilizing scanning electron microscopy (SEM)

and also investigating the influence of triangular wave shape pulsed current on improved corrosion resistance of nickel metal matrix with incorporated ZrO_2–TiO_2 nanocomposite on the mild steel substrate. The results of the pitting corrosion measurements were corroborated with surface morphology and X-ray diffraction patterns.

5.2 Procedure

5.2.1 Materials and Synthesis of Nanocomposites

TiO_2 [avra-98%], $ZrO_2Cl_2{\cdot}8H_2O$ [Himedia-99%], isopropanol [PA-CH_3CH (OH) CH_3, FINAR], NH_4OH pellets, HCl [≥35% Merck. Himedia], distilled water, and cetyl trimethyl ammonium bromide (CTAB) chemicals were used to synthesize 50% ZrO_2–TiO_2 nanocomposites.

In this work, sol-gel method was used to synthesize nanostructured ZrO_2–TiO_2 composites, and the synthesis procedure is shown in Figure 5.1(a). First, two solutions were prepared: (1) 10 g of titanium dioxide (TiO_2) was dissolved in 9.3 mL of isopropanol and sonicated for 3 min at room temperature and stirred for 2 h with a magnetic stir bar at 70°C temperature, and (2) 2.615 g of zirchonyl chloride ($ZrOCl_2{\cdot}8H_2O$) was dissolved in 100 mL distilled water. Solution 2 was added dropwise to solution 1 under

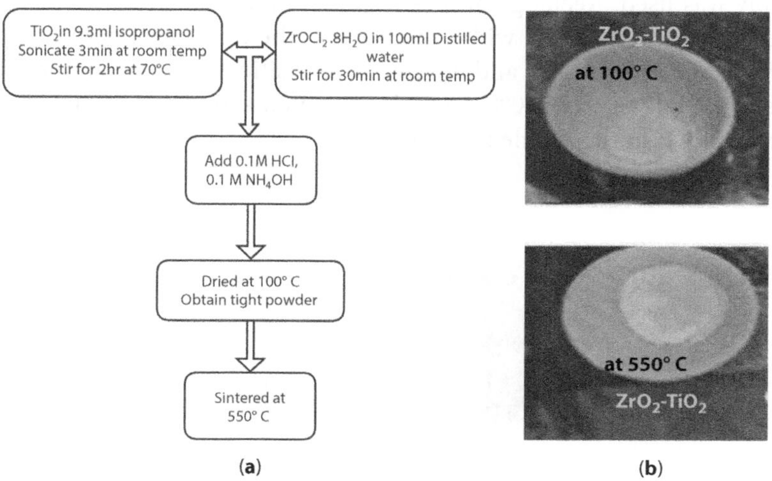

Figure 5.1 (a) Synthesis procedure of 50% ZrO_2–TiO_2 nanocomposites. (b) Prepared powder.

vigorous stirring, and they were mixed together thoroughly. Moreover, 0.125 g of CTAB was added; it acts as a surfactant to the resulting solution and also for the purpose of particle size reduction. Furthermore, the solution was acidified with 0.1 M HCl, and 0.1 M NH_4OH was added to adjust the pH value of 1.3. The experiment was done under continuous stirring of 550 rpm for 2 h at 70°C temperature. Afterward, the prepared gel was dried at 100°C with a heating mantle and then sintered for 2 h at 550°C in a muffle furnace. Finally, 50% ZrO_2–TiO_2 nanocomposites were obtained, and the prepared powder is shown in Figure 5.1(b). Then, the powder was analyzed using SEM and XRD.

5.2.2 Electrodeposition

Ni–nano ZrO_2–TiO_2 composite coatings were applied on mild steel substrates using a two-electrode system, employing both rectangular and triangular current pulses. The mild steel substrates underwent a stepwise grinding process with 320, 800, 1/0, 2/0, and 3/0 emery papers, followed by metallographic polishing until a mirror-bright surface finish was achieved. Subsequently, the samples were ultrasonically cleaned using distilled water and acetone in 1:1 ratio and then air-dried prior to the deposition process. The prepared sample is depicted in Figure 5.2. The elemental analysis of the cathode sample is detailed in Table 5.1. The experimental setup was constructed for laboratory investigation as shown in Figure 5.3. In the electrodeposition process, GW INSTEK 3030D regulated direct current power supply was used. Rectangular and triangular periodic pulses were generated with Arduino pulse width modulator. The electrolyte bath composition is given in Table 5.2, and the electrodeposition process parameters are given in Table 5.3. Ni–50% ZrO_2–TiO_2-coated samples were annealed at 700°C for 2 h in the muffle furnace.

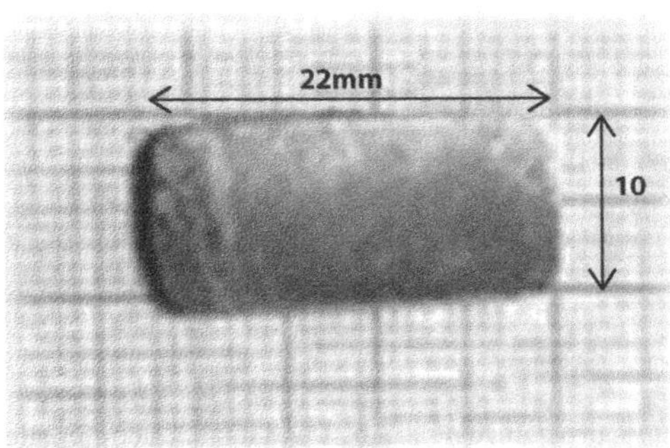

Figure 5.2 Mild steel substrate.

Table 5.1 Elemental analysis of mild steel (in wt.%).

Elements	C	Mn	Si	S	P	Ni	O	Ferrous
Mild steel	0.98	0.39	0.08	0.07	0.05	0.17	0.08	Balance

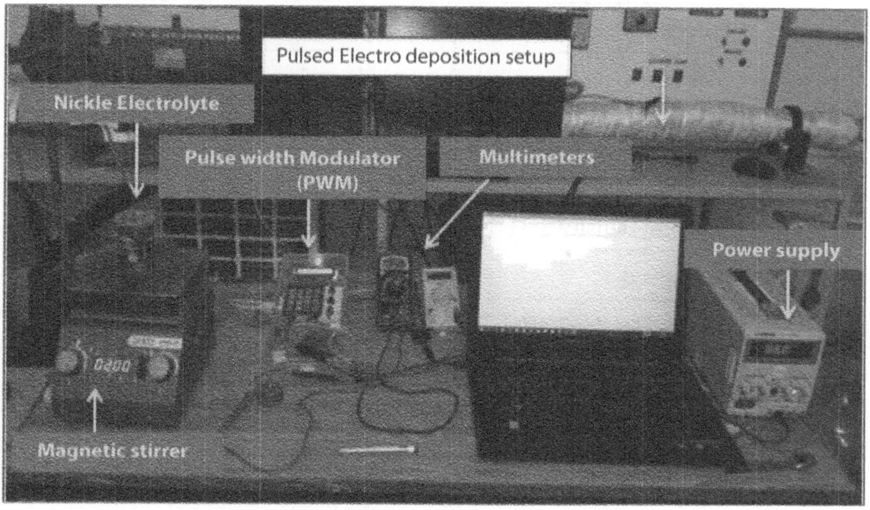

Figure 5.3 Electrodeposition experimental setup.

Table 5.2 Composition of nickel electrolyte.

Materials	Quantity
$NiSO_4{\cdot}H_2O$	260 g/L
$NiCl_2{\cdot}6H_2O$	60 g/L
Boric acid	38 g/L
CTAB (cetyl trimethyl ammonium bromide)	0.5 g/L
Polyethylene glycol	1 mL/L
50% ZrO_2–TiO_2	3 g/L
Anode	Nickel
Cathode	Mild steel
NaOH	0.1 M
HCl	0.1 M

Table 5.3 Electrodeposition conditions.

Parameter	Value
Temperature	Room temp. (30°C)
Plating time	15 min
frequency	10 Hz
Waveform	Rectangular, triangular pulses
Duty cycle	10%
Voltage	250 mV
Current density	10 mA/cm^2
Stir speed	250 rpm
Electrodes Separation	3 cm
pH	4.1

5.2.3 Characterization

Diffraction analysis utilized CuKα (λ = 0.15406 nm) in a Pananalytical X-ray diffractometer to examine the phase and grain sizes of the deposits. The surface morphology of the prepared powder and deposits was observed through scanning electronic microscopy (SEM) analysis [18, 19]. A Vickers microhardness tester was used to determine the surface strength of the nickel nanocomposite deposits. The microhardness of the coatings was assessed using an indentation technique at 50 gf load applied and dwell time of 10 s as per ASTM E384 standard [20–23]. Corrosion resistance assessment of deposits was conducted *via* potentiodynamic polarization test by employing a 3.5%-NaCl solution with a three-electrode cell system. The tests utilized a Ag/AgCl electrode as the reference electrode, a platinum plate as counter electrode, and an electrodeposited sample as working electrode [24, 25].

5.3 Results and Discussions

5.3.1 Microscopic Structure Analysis

Figure 5.4 depicts the SEM morphology of the ZrO_2–TiO_2 particles. The particles appear predominantly spherical and are observed to be below 100 nm

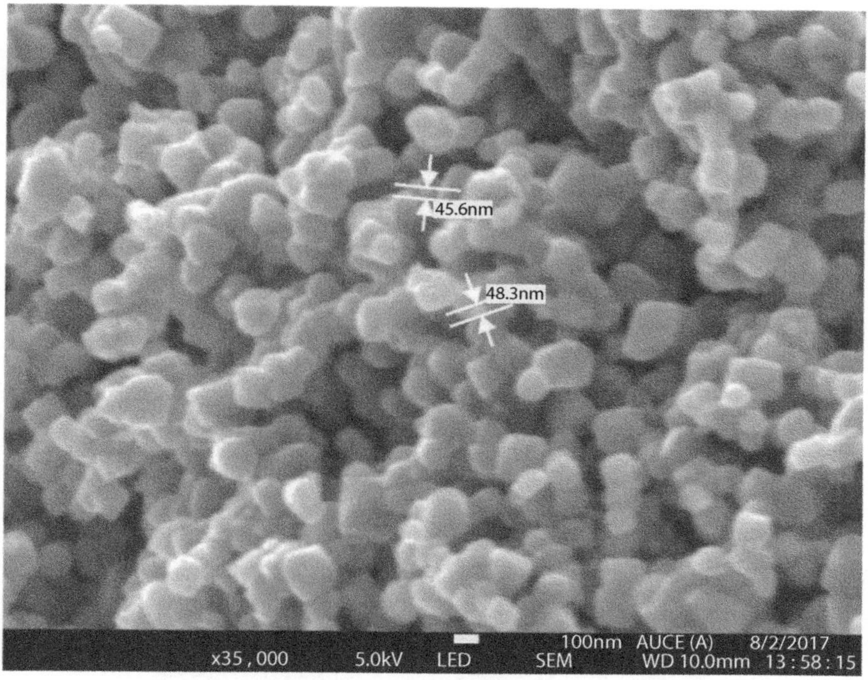

Figure 5.4 Surface micrographs of nano-ZrO_2–TiO_2 powder.

in size. Uncoated mild steel surface is shown in Figure 5.5(a). The surface of Ni/50% ZrO_2–TiO_2 nanocomposite coatings on mild steel deposited under rectangular and triangle pulses at 10-Hz frequency and 10% duty cycle is shown in Figure 5.5(b, c). Ni–50% ZrO_2–TiO_2 nanocomposite layer coating with both pulses produces a smoother and more uniform surface. Both rectangular and triangular waveforms with 10% relaxation time significantly influence the surface morphology of nanocomposite coatings. More agglomeration of composites was observed in the rectangular pulsed coatings as shown in Figure 5.5(b). During the growth process of metal coatings, various textures can emerge. This occurrence stems from discrepancies in surface energy, leading to the selective growth of grains with the lowest surface free energy, but the rectangular pulses produced more pore density and flake structure, whereas reduced pore density and finer grain structures were formed with triangular pulsed currents as shown in Figure 5.5(c). It yields a smaller grain size than that of the rectangular wave form. The diffusion current density is low with triangular pulse current wave form, and it is detrimental to crystal growth. It led to create new nuclei and suppressed the grain growth.

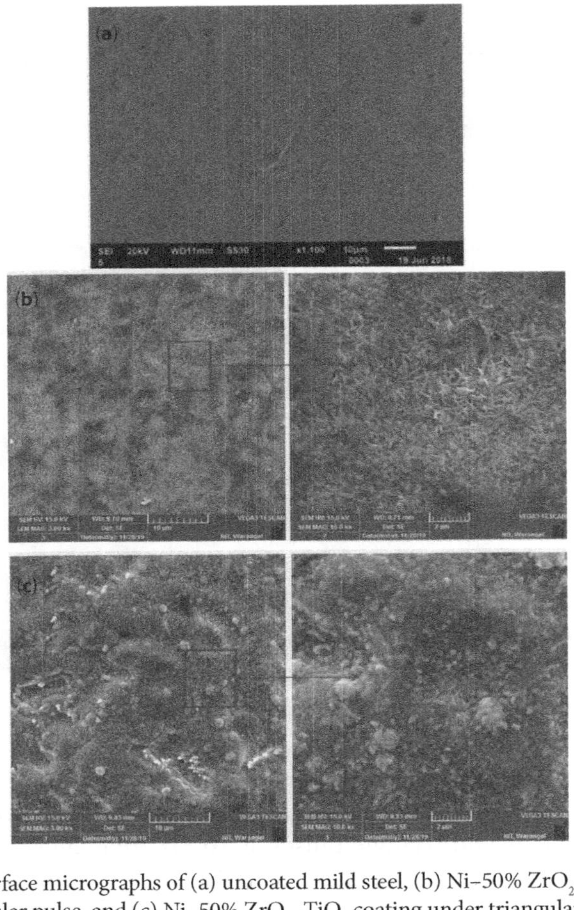

Figure 5.5 Surface micrographs of (a) uncoated mild steel, (b) Ni–50% ZrO_2–TiO_2 coating under rectangular pulse, and (c) Ni–50% ZrO_2–TiO_2 coating under triangular pulse.

5.3.2 X-Ray Diffraction

Figure 5.6 shows the XRD pattern of as-prepared ZrO_2–TiO_2 binary oxide powder. The diffraction peaks were matched with JCPDS data card no: 01-080-1783 of zirconium titanium oxide, and it indicates orthorhombic crystal structure. The observed crystalline size of particles was 20 nm using Debye–Scherrer equation. The absence of significant broadening in the XRD peak, despite the ultrafine nature of the particles, suggests that the broadening primarily stems from grain refinement rather than from strain induced by the chemical route used in nanocomposite powder preparation.

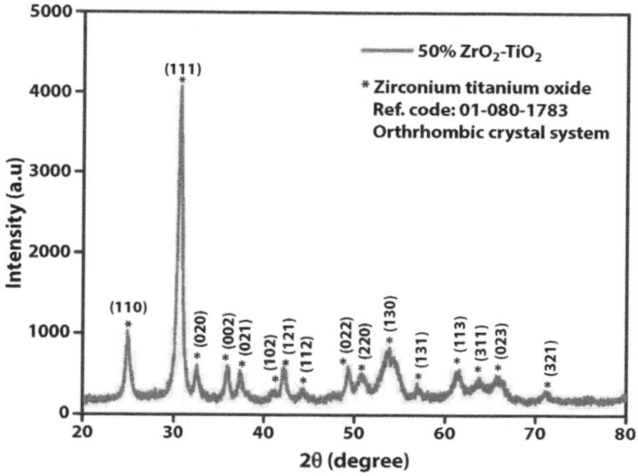

Figure 5.6 XRD pattern of nano-ZrO_2–TiO_2 powder.

It was confirmed that this structure is orthorhombic, and a strong (111) peak expansion was observed. This may be due to the additional effect of zirconium atoms, which reduce the crystalline size of the titanium oxide particles. Figure 5.7 shows the diffraction peaks of uncoated and Ni deposits with the addition of ZrO_2–TiO_2 ceramic particles, produced using rectangular and triangular shape pulsed current electrodeposition process. Mild steel substrate diffraction peaks at $2\Theta = 44.82$, 65.05, and 82.3 were observed. The XRD patterns of deposits with rectangular pulse showed peaks at $2\Theta = 35.76$, 43.25, 49.55, 54.21 (Ni), 57.09, 62.67, 64.03, 74.46 (Ni) and for the triangular pulse showed at $2\Theta = 35.84$, 41.09, 43.30, 49.65, 54.33 (Ni), 57.13, 62.66, 64.11, 75.51 (Ni). The coatings completely suppressed the substrate peaks. In the two co-deposited samples, there was small significant intensity of ZrO_2–TiO_2 peaks, i.e., deposits with triangular pulse enlarged the ZrO_2–TiO_2 intensity peaks with (002) crystal plane. Large amounts of particles may act as inhibitors of (111) growth with the triangular pulse. From the abovementioned observation, it can be said that more zirconia–titania composites were accumulated toward cathodic mild steel with rectangular pulse frequency, with both rectangular- and triangular-pulsed electrodeposited coatings consisting of high ZrO_2–TiO_2 composites and lower nickel content.

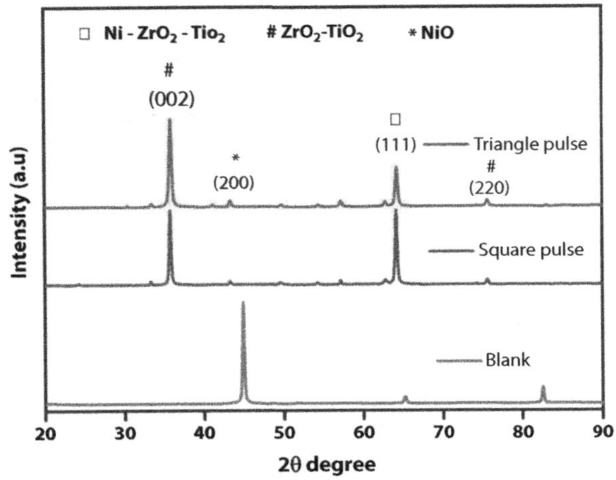

Figure 5.7 XRD pattern of blank, Ni/nano-ZrO_2–TiO_2 coating using rectangular and triangular pulse current.

5.3.3 Microhardness

The microhardness of Ni deposits with nano ZrO_2–TiO_2 composites under rectangular and triangular waveforms are given in Table 5.4. The smaller grain size was obtained at triangular pulsed current with 10% duty cycle and 10 Hz lower frequency when compared to rectangular pulsed current with the same deposition conditions. In the electrodeposition process, the grain size is related to the nucleation rate. At a high overpotential during the rectangular pulse, there was an observed increased in nucleation rate with fine grain size. The diffusion time of nickel ions on the surface is significantly reduced, so they do not have time to find lower energy centers that contribute to their crystal growth. However, with the triangular waveform, the enhancement in hardness could be attributed to the smooth and lower grain size of 75 nm of the deposits. It favors nucleation rather than

Table 5.4 Microhardness of Ni deposits with ZrO_2–TiO_2 under rectangular and triangular waveforms.

Sample	Microhardness (Hv)
Base mild steel	342 ± 9
Coating with rectangular wave	784 ± 15
Coating with triangular wave	874 ± 19

crystal growth. It led to a rise in coating hardness that correlates with the reduction in grain size, which adheres to the Hall–Petch relation. Therefore, in the process of electrodeposition, this mechanism can be precisely implemented in the form of a triangular wave with a shorter pulse on-time.

5.3.4 Pitting Corrosion Studies

Electrochemical tests were carried out using a conventional three-electrode cell at 3.5% NaCl solution at room temperature. The Tafel curves of uncoated and coated steels are presented in Figure 5.8. Comparing the coated mild steel with the uncoated mild steel specimens results in the remarkable displacement of the Tafel curves toward the lower current densities and corrosion potential toward positive values. The corrosion parameters such as potential, current density, and protection efficiency values were calculated and are presented in Table 5.5. Pure nickel coatings above -690 mV (vs. SCE) current density begin to increase gradually with increasing potential. The addition of ZrO_2–TiO_2 composites in the nickel electrolyte system may influence the nucleation and growth mechanism of nickel deposits *via* electrodeposition. The applied electric field and added nanocomposites in Ni electrolyte encouraged the Ni along with nanoparticle transportation toward the cathode surface and are strongly adsorbed. It activates the surface and produces high nucleation sites for Ni^{2+} ion reduction.

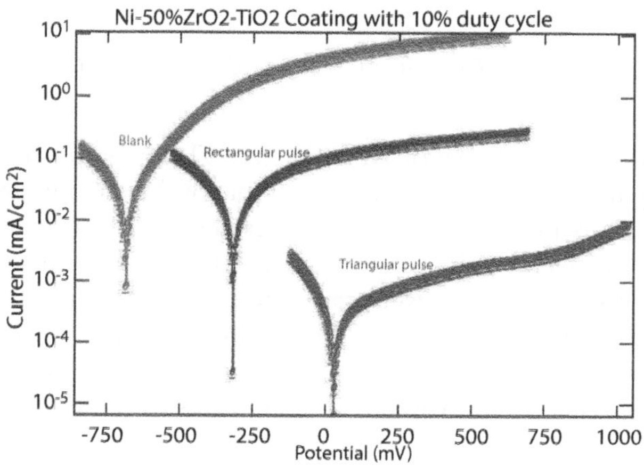

Figure 5.8 Tafel plots of blank and Ni/nano-ZrO_2–TiO_2 coating using rectangular and triangular pulse current.

Table 5.5 Electrochemical parameters of Tafel curves of Ni–50% ZrO_2–TiO_2 nanocomposite coating.

Sample name	βa mV	βc mV	Ecorr mv	Icorr mA/cm²	Corr. rate mm/year	LPR ohm/cm²	Protection efficiency PF%
Uncoated	25.01	27.08	-699	4.9×10^{-02}	5.7×10^{-01}	1.14×10^{2}	---
Rectangular pulse	18.7	22.2	-321	1.2×10^{-03}	1.4×10^{-02}	3.4×10^{3}	97.49
Triangle pulse	39.94	31	56	3.2×10^{-05}	3.7×10^{-04}	2.4×10^{5}	99.93

The cathodic reactions 5.1 and 5.2 represent the pure nickel and 5.3, 5.4, and 5.5 represent the cathodic Ni reduction by the effect of ZrO_2–TiO_2 nanocomposites on the nickel electrodeposited coating.

$$Ni^{2+} + e^{-} \rightarrow Ni^{+}_{ads} \quad (5.1)$$

$$Ni^{+}_{ads} + e^{-} \rightarrow Ni \quad (5.2)$$

$$Ni^{2+} + TiO_2 + ZrO_2 \rightarrow Ni^{2+}_{ads}(ZrO_2\text{-}TiO_2) \quad (5.3)$$

$$Ni^{2+}_{ads}(ZrO_2\text{-}TiO_2) + e^{-} \rightarrow Ni^{+}_{ads}(ZrO_2\text{-}TiO_2) \quad (5.4)$$

$$Ni^{+}_{ads}(ZrO_2\text{-}TiO_2) + e^{-} \rightarrow Ni / ZrO_2\text{-}TiO_2 \quad (5.5)$$

The corrosion resistance of Ni–50% ZrO_2–TiO_2 composites under the triangular pulse frequency is greater than that of the rectangular pulse frequency, which is in accordance with the triangular pulse average current density that is lower, produced lower pore density, and with a smooth and homogenous surface with lower grain size than rectangular pulsed deposition. The corrosion resistance of composite coatings is to be influenced by the homogeneity of the layer, pore size and density, smoothness, matrix grain size as well as reinforcement of composites with metal coatings.

5.4 Conclusions

Nickel/ZrO_2–TiO_2 nanocomposite coatings were made on mild steel specimens using rectangular and triangular waveform current electrodeposition at 10% duty cycle and 10-Hz frequency and were heat-treated. The triangular wave shape pulsed current produced a uniform and large amount of 50% ZrO_2–TiO_2 particles with a smaller grain size in Ni metal matrix deposits than the deposits under the rectangular pulsed deposition process. The deposits under the triangular pulse obtained the highest protection efficiency from the 3.5% NaCl corrosive environment than that of the rectangular pulse. Finer grains, smooth and uniform layers, and less pore density are the main causes to enhance the corrosion resistance and microhardness of deposits made by triangular waves.

References

1. Prasad, C., Bhavani, K., Vemuri, R., Ramana, V.V., Sriramvikas, K., Effect of duty cycle variation on Nickel electrodeposits at 10 Hz frequency. *Chem. Data Collect.*, 41, 100927, 2022, https://doi.org/10.1016/j.cdc.2022.100927.
2. Prasad, C., Srinivasa Rao, K., Ramji, K., Studies on Pitting Corrosion of Pulsed Electrodeposited Nanocomposite Coating, pp. 404–412, 2020, https://doi.org/10.1007/978-3-030-24314-2_49.
3. Sri Ram Vikas, K., Rahul, Kishore Reddy, C., Venkata Ramana, V.S.N., Morampudi, P., Prasad, C., Effect of Grade 5 titanium interlayer on microstructure and pitting corrosion behaviour of AA1100/A36 explosion welds. *Mater. Today Proc.*, 59, 1659–1665, 2022, https://doi.org/10.1016/j.matpr.2022.03.330.
4. Eswara, J., Raju, B.N., Prasad, C., Sankar, B.S.S.P., Machining performance on SS304 using nontoxic, biodegradable vegetable-based cutting fluids. *Chem. Data Collect.*, 42, 100961, 2022. https://doi.org/10.1016/j.cdc.2022.100961.
5. Arango, S. and G. Group, Coating and Surface Treatments on Orthodontic Metallic Materials. *Coatings*, 3, 1, 1–15, 2013, https://doi.org/10.3390/coatings3010001.
6. Andersons, J. and Leterrier, Y., Advanced fragmentation stage of oxide coating on polymer substrate under biaxial tension. *Thin Solid Films*, 471, 209–217, 2005, https://doi.org/10.1016/j.tsf.2004.07.007.
7. Bhardwaj, M., Balani, K., Balasubramaniam, R., Pandey, S., Agarwal, A., Effect of current density and grain refining agents on pulsed electrodeposition of nanocrystalline nickel. *Surf. Eng.*, 27, 642–648, 2011, https://doi.org/10.1179/026708410X12683118611185.

8. Cheraghi, H., Shahmiri, M., Sadeghian, Z., Corrosion behavior of TiO_2-NiO nanocomposite thin films on AISI 316L stainless steel prepared by sol-gel method. *Thin Solid Films*, 522, 289–296, 2012, https://doi.org/10.1016/j.tsf.2012.07.125.
9. Saucedo, E., Corregidor, V., Fornaro, L., CuÃ±a, A., Dieguez, E., CdTe polycrystalline films for X-ray digital imaging applications. *Thin Solid Films*, 471, 304–309, 2005, https://doi.org/10.1016/j.tsf.2004.06.091.
10. Saladi, S., Menghani, J.V., Prakash, S., Characterization and Evaluation of Cyclic Hot Corrosion Resistance of Detonation-Gun Sprayed Ni-5Al Coatings on Inconel-718. *J. Therm. Spray Technol.*, 24, 778–788, 2015, https://doi.org/10.1007/s11666-015-0235-1.
11. Li, Y., Jiang, H., Huang, W., Tian, H., Effects of peak current density on the mechanical properties of nanocrystalline Ni-Co alloys produced by pulse electrodeposition. *Appl. Surf. Sci.*, 254, 6865–6869, 2008, https://doi.org/10.1016/j.apsusc.2008.04.087.
12. Wasekar, N.P., Haridoss, P., Seshadri, S.K., Sundararajan, G., Influence of mode of electrodeposition, current density and saccharin on the microstructure and hardness of electrodeposited nanocrystalline nickel coatings. *Surf. Coat. Technol.*, 291, 130–140, 2016, https://doi.org/10.1016/j.surfcoat.2016.02.024.
13. Maharana, H.S., Basu, A., Mondal, K., Structural and tribological correlation of electrodeposited solid lubricating Ni-WSe_2 composite coating. *Surf. Coat. Technol.*, 349, 328–339, 2018, https://doi.org/10.1016/j.surfcoat.2018.06.005.
14. Li, B., Zhang, W., Li, D., Wang, J., Electrodeposition of Ni[sbnd]W/ZrO_2 nanocrystalline film reinforced by CeO_2 nanoparticles: Structure, surface properties and corrosion resistance. *Mater. Chem. Phys.*, 229, 495–507, 2019, https://doi.org/10.1016/j.matchemphys.2019.03.049.
15. Choi, W.H., Kwak, B.S., Noh, Y.H., Yook, J.G., Kweon, J.H., Nam, Y.W., Radar-absorbing nickel-coated fabric composite for wing-shaped structure in the X-band. *Compos. Struct.*, 239, 111885, 2020, https://doi.org/10.1016/j.compstruct.2020.111885.
16. Nagasree, P.S., Vandana, K.I.V., Hima, C., Naidu, M.K., Prasad, C., Integrated MCDM techniques for the selection of hybrid MWCNT / Ni – Zn – Fe nanocomposites for X-band radar absorption applications. *Inor. Chem. Commun.*, 161, 112139, 2024.
17. Ullal, Y. and Chitharanjan Hegde, A., Corrosion protection of electrodeposited multilayer nanocomposite Zn-Ni-SiO_2 coatings. *Surf. Eng. Appl. Electrochem.*, 49, 161–167, 2013, https://doi.org/10.3103/S1068375513020142.
18. Eswara Manikanta, J., Kumar Gurajala, N., Prasad, C., Santha Rao, D., Srinivasa Vara Prasad, A., Phani Sankar, B., Section A-Research paper Investigational Study on Influencing Parameters for Flexural Strength and Hardness of NITINOL-Based Aluminum Matrix Composites. *Chem. Bull.*, 12, 3, 683–691, 2023, https://doi.org/10.31838/ecb/2023.12.s3.079.

19. Prasad, C., Jagadeesh, K., Sreenivasa Rao, P., Sastry, D.L., Jose, S., Vijaya Babu, K., Synthesis and Characterization of Ag-Cds Nanostructures. *Int. J. Adv. Eng. Technol.*, 9, 84–90, 2016.
20. Koona, B., Venkata Ramana, V.S.N., Prasad, C., Rahul, Sri Ram Vikas, K., Comparison of microstructure and corrosion behaviour of AA2014 electron beam and friction stir welds. *Mater. Today Proc.*, 52, 1615–1621, 2022, https://doi.org/10.1016/j.matpr.2021.11.272.
21. Morampudi, P., V.R. V.S.N., Chitrada, P., S.V.K., R.R., Tribology and pitting corrosion behavior of Al6061/nano-ZrB2 metal matrix composites prepared *via* powder metallurgy process. *Chem. Data Collect.*, 48, 101082, 2023, https://doi.org/10.1016/j.cdc.2023.101082.
22. Morampudi, P., Ramana, V.S.N.V., Vikas, K.S.R., Rahul, R., Prasad, C., Effect of nano ZrB2 particles on physical, mechanical and corrosion properties of Al6061 metal-matrix nano composites through stir casting route. *Eng. Res. Express*, 4, 2, 025010, 2022, https://doi.org/10.1088/2631-8695/ac5f66.
23. Reddy, C.K., Prasad, C., Vikas, K.S.R., Reddy, M.S., Rahul, Microstructural and Mechanical Properties of Friction Stir Welded (FSW) AA7075 with B C Reinforcement, in: *Recent Advances in Mechanical Engineering*, vol. 2, ICMech-REC 2023. Lecture Notes in Mechanical Engineering, Springer, Singapore, 2024, https://doi.org/10.1007/978-981-97-2249-5_28.
24. Morampudi, P., Venkata Ramana, V.S.N., Vikas, K.S., Rahul, Prasad, C., Enhancing wear properties of Al6061 metal-matrix composites by reinforcement of ZrB2 nano particles. *Mater. Today Proc.*, 59, A45–A51, 2022, https://doi.org/10.1016/j.matpr.2022.04.865.
25. Morampudi, P., Venkata Ramana, V.S.N., Prasad, C., SriramVikas, K., Rahul, Physical, mechanical and corrosion properties of Al6061/ZrB2 metal matrix nano composites *via* powder metallurgy process. *Mater. Today Proc.*, 59, 1708–1713, 2022, https://doi.org/10.1016/j.matpr.2022.03.596.

6

Smart Composite Materials for Aerospace Applications

Ibrahim Momoh-Bello Omiogbemi[1]*, Emmanuel Imhanote Awode[1], Mohammed Habib Muhammad[1], Adegoke Adesanmi[2] and Ishaya Musa Dagwa[3]

[1]*Mechanical Engineering Department, Air Force Institute of Technology, Kaduna, Nigeria*
[2]*Mechanical Engineering Department, University of Jos, Jos, Nigeria*
[3]*Mechanical Engineering Department, University of Abuja, Gwagwalada, Abuja, Nigeria*

Abstract

This chapter delves into the advancing field of smart functional composite polymer materials designed for aerospace applications. The discussion starts by emphasizing the pivotal importance of composite materials in the aerospace industry, particularly their lightweight and high-strength characteristics that boost performance and fuel efficiency. It categorizes and explains various smart functional composites, such as polymer matrix composites (PMCs), carbon fiber-reinforced polymers (CFRPs), glass fiber-reinforced polymers (GFRPs), shape memory polymers (SMPs), and self-healing polymers. Their properties are analyzed in depth, covering mechanical, thermal, electrical, and multifunctional characteristics. The chapter further explores the wide-ranging applications of these materials in aerospace, such as their use in structural components, functional devices, thermal management systems, and self-repairing coatings. A review of manufacturing techniques, from traditional methods to advanced technologies like 3D printing and nanotechnology-driven production, is also provided. The challenges associated with integrating smart composites into existing aerospace systems are discussed based on existing literature, along with future research opportunities aimed at improving scalability and cost efficiency. Ultimately, the chapter emphasizes the transformative potential of smart functional composite polymer materials in driving innovation and optimizing performance in aerospace technology.

**Corresponding author*: omiogbemi1@gmail.com

Sandip Kunar, Pranav Charkha, Santosh Jaju and Harish Tiwari (eds.) Functional Composites: Role in Modern Engineering, (115–168) © 2025 Scrivener Publishing LLC

Keywords: Smart functional composites polymers, PMCs, CFRPs, GFRPs, SMPs, self-healing polymers, nanotechnology-driven production, aircraft technologies

6.1 Introduction

The rapid advancement of products, technologies, and industrial practices is driving the development of new materials designed to manage and stabilize internal structures against high stresses, deformations, crack propagation, and environmental challenges. Among these materials, composites are continually evolving to produce options that are either cost-effective, high-performance, or both. Additionally, there is a growing emphasis on environmental protection and public health, leading to the development of eco-friendly composites. Notably, "smart" materials have emerged as significant players in manufacturing and technology. These materials are embedded within other structures to monitor their health and detect weaknesses or potential failures. They can respond appropriately by signaling necessary modifications or initiating corrective actions to prevent total deformation or failure. Today's technological demands necessitate the creation of such materials to ensure dimensional stability and control in modern structures, including high-rise buildings and aerospace applications, ultimately protecting the environment from disasters and diseases [1]. Smart materials are often composite materials designed to perform multiple functions. Their core concept involves integrating additional capabilities to transform basic structural materials into functional, active, and adaptable systems that can be controlled or directed, even from significant distances. Achieving this requires the inclusion of specific mechanisms or materials to ensure precise actuation, dimensional stability, and autonomy. This also requires extensive experimentation and validations before deployment to the field.

Therefore, the goal of these materials extends beyond performing programmed functions; they aim to learn and respond appropriately to a wide range of situations. Technologically, this is realized through the integration of several macroscopically distinct components around a primary base, typically a composite material [2, 3]. Two key characteristics are commonly observed in these materials [1–3]:

1. Sensitivity: The ability of the material to provide information about its environment or its own state.

2. Adaptability: The capability to produce actions that modify the system's behavior in response to changes.

A material, system, or structure possessing both sensitivity and adaptability is termed adaptive, meaning, it collects information, processes it, and reacts appropriately. However, adaptive systems respond to stimuli based on predetermined laws defined during development and lack the ability to choose among multiple responses. In contrast, smart materials are not only sensitive and adaptive but also evolutionary, capable of selecting responses from a range of possibilities. They act as sensors (detecting signals), actuators (interacting with the environment), and processors (handling data). These materials spontaneously modify properties like shape, conductivity, viscoelasticity, or color in response to stimuli such as temperature changes, stress, or electromagnetic fields. Smart materials can detect structural weaknesses, crack propagation, or vibrations, enabling solutions like aircraft coatings or vibration reduction. Shape memory alloys (SMAs) and piezoelectric materials exemplify this, offering unique thermo-mechanical or electromechanical properties for advanced applications [4, 5]. Therefore, smart composite materials, often termed adaptive or intelligent composites, are advanced systems engineered to respond predictably to environmental stimuli. They form the functional backbone of intelligent designs, providing dynamic capabilities such as self-monitoring, self-repair, and active interaction with their surroundings [6–8]. These materials, integrated with artificial intelligent (AI) and machine learning, have diverse applications in fields such as coatings, energy, manufacturing, and cybersecurity. Their multifunctionality and adaptability make them essential for emerging technologies like additive manufacturing, where they perform specific tasks and autonomously adapt to conditions. Smart materials, including piezoelectric, shape memory alloys, hydrogels, and self-healing polymers, simplify system complexity and enhance implementation. Traditional composites combine polymer, ceramic, or metal matrices with reinforcements like fibers for strength, while smart composites innovate with active responsiveness [9–12]. Research has focused on leveraging adaptive control and smart materials for improved performance in aerospace and other industries [6, 8].

In aerospace, the use of smart structures remains limited for aeroelastic lifting surfaces and vibration suppression in structural components. Innovations like smart airfoils, which induce twist and camber in wings to generate specific aerodynamic forces and moments, eliminate the need for conventional control surfaces, enabling advanced aeroelastic control [6, 7, 13].

Investigations have explored applications in helicopter rotor blades, aircraft wings, and missile fins, demonstrating their potential for vibration control, flutter suppression, and flight performance optimization. Beyond aerospace, smart structures are increasingly applied in automobile engineering, construction, and strategic defense systems, such as enhanced target tracking, health monitoring, and threat assessment technologies [13, 14]. Polymer composites, including smart variants, are widely valued across industries for their exceptional durability, lightweight nature, cost-effectiveness, adaptability, and environmentally friendly properties [15, 16]. The aerospace industry requires materials that combine lightweight properties, high strength, durability, and multifunctionality. Smart functional composite polymer materials fulfill these requirements by integrating advanced polymer matrices with functional reinforcements, providing innovative solutions for aerospace applications [2, 3, 5, 11, 12, 17].

6.1.1 Importance of Composite Materials in Aerospace

Because of their special qualities and benefits over conventional materials, composite materials have become indispensable in aeronautical engineering. These materials are made by mixing two or more components, usually reinforcements (like fibers or particles) and a matrix (like metal, ceramic, or polymer), which results in structures that are strong, lightweight, and extremely durable [18–21]. Figure 6.1 illustrates the schematic diagram of composite material structure. Their significance in aerospace applications can be highlighted in several key areas [22].

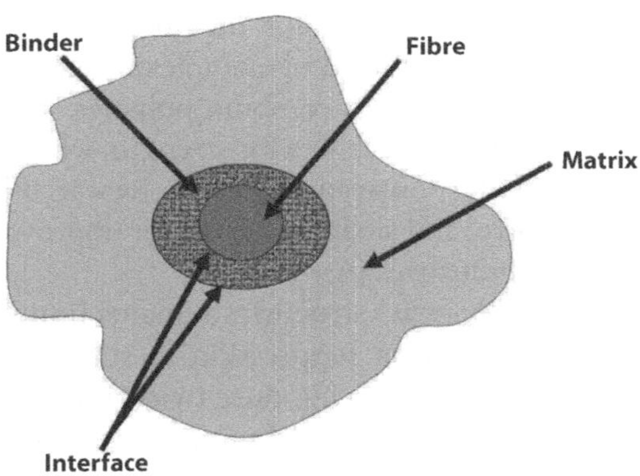

Figure 6.1 Schematic diagram of a typical composite material structure.

6.1.1.1 Weight Reduction

The high weight-to-strength ratio, also referred to as "specific strength," is a fundamental property of composite materials. The tensile or compressive force per unit area at the moment of failure, divided by the density of the material, is known as specific strength. The strength-to-weight ratio is a different designation for this ratio. The common metric for specific strength in applications using composite fiber laminates is usually tenacity. By reducing the weight of aircraft structures, significant operational cost savings can be achieved, leading to increased payload capacity, lower maintenance expenses, and an extended flight range, along with numerous other benefits [18, 19]. Composite materials are considerably lighter than metals such as steel and aluminum while still providing high strength. This reduction in weight leads to improved fuel efficiency, decreased emissions, and increased payload capacity in both aircraft and spacecraft [23, 24].

6.1.1.2 High Strength-to-Weight Ratio

Aerospace structures, such as fuselages, wings, and rotor blades, must be built from materials capable of withstanding extreme forces and challenging environmental conditions. In this context, composite materials provide an ideal solution due to their exceptional strength-to-weight ratio. This characteristic not only contributes to improved structural integrity but also enhances overall performance and safety in aerospace applications. Composites are engineered to deliver high tensile strength while remaining lightweight, which is essential for optimizing fuel efficiency and payload capacity in aircraft. Furthermore, their resistance to fatigue and environmental degradation ensures that these structures can maintain their performance over time, even in harsh operating conditions. As the aerospace industry continues to evolve, the adoption of composite materials is becoming increasingly vital for developing advanced aircraft that meet stringent safety standards while achieving greater efficiency and sustainability. Overall, composites play a crucial role in advancing aerospace technology and enhancing the capabilities of modern aircraft [19, 23, 24].

6.1.1.3 Corrosion Resistance

Composite materials' practically insignificant corrosion susceptibility is one of its distinctive characteristics. Composites have exceptional resistance to galvanic corrosion, which is a major reason for structural repairs in aluminum aircraft, in contrast to metals, especially aluminum. By using composite materials, designers can create structures with almost limitless operational fatigue life, such as wings, fuselages, flight controls, turbofan engine blades, and even helicopter rotor blades. In addition to extending the lifespan of aeronautical parts, this resilience lowers maintenance expenses and raises general safety. The ability to design corrosion-resistant structures significantly contributes to the efficiency and reliability of modern aircraft, making composites an increasingly preferred choice in aerospace engineering [25, 26]. In contrast to metals, composites are resistant to corrosion and fatigue when exposed to harsh environmental conditions like moisture, UV radiation, and salt spray. This property helps lower maintenance costs and prolongs the lifespan of aerospace components [27].

6.1.1.4 *Design Flexibility*

Composites can be shaped into complex forms, enabling designers to enhance aerodynamic performance and integrate multiple functions within a single component. This versatility is crucial for modern aircraft design, as it allows for streamlined structures that improve efficiency and reduce weight. By utilizing composite materials, engineers can create innovative solutions that fulfill the exacting standards of the aerospace sector. The ability to mold composites into specific shapes not only optimizes airflow but also facilitates the incorporation of various systems, such as control surfaces and structural elements, into one cohesive unit. This design flexibility contributes significantly to the performance, safety, and overall effectiveness of contemporary aircraft, making composites an invaluable asset in aerospace engineering [28–30].

6.1.1.5 *Heat Resistance and Thermal Stability*

During flight, aircraft structures experience significant and rapid temperature fluctuations. For instance, in hot climates, an aircraft may begin on the ground with surface temperatures exceeding 120°F, only to encounter ambient air temperatures dropping to -60°F or lower as it ascends. Structural thermal stability is crucial in aircraft design. Unlike the all-metal supersonic Concorde, which would stretch due to air friction heating, the new all-composite Boom Supersonic passenger airliner is designed to maintain its integrity under these conditions [23, 31, 32]. Composite materials, particularly those reinforced with glass and carbon fibers, demonstrate exceptional resistance to degradation under extreme high-temperature conditions. These properties have made them indispensable in aerospace engineering, especially in thermal protection systems. For example, ablative heat shields, used in the Mercury, Gemini, and Apollo missions, were critical for withstanding the intense heat generated during atmospheric re-entry. Similarly, the Space Shuttle's thermal protection tiles showcased the resilience of ceramic-coated composites. In the realm of advanced aerospace technology, carbon-based and ceramic-reinforced composites are pivotal for developing hypersonic passenger aircraft that operate at speeds exceeding Mach 6. At such velocities, the intense atmospheric friction causes outer surface temperatures to rise beyond 1,000°F. These conditions necessitate materials that not only endure extreme heat but also maintain structural integrity and performance, making advanced composites the cornerstone of future hypersonic aerospace applications [23, 33]. Composites can be tailored to possess specific thermal and electrical characteristics, making them ideal for applications such as heat shielding, insulation, and radar absorption in both military and commercial aerospace sectors [32].

6.1.1.6 Structural Health Monitoring

Advanced composites that incorporate smart materials, such as piezoelectric fibers, facilitate real-time monitoring of structural health. This innovative capability significantly enhances the safety and reliability of aerospace structures by providing continuous feedback on their condition. By detecting potential issues early, these smart composites allow for timely interventions, ultimately reducing the time needed for inspections and repairs. This proactive approach not only minimizes downtime but also helps prevent costly failures, ensuring that aircraft remain in optimal operating condition. Thus, the incorporation of smart materials into composite structures represents a major advancement in aerospace engineering, contributing to improved performance and operational efficiency in modern aviation [34–36].

6.1.1.7 Reduction in Production Costs

Modern methods of production like automated fiber placement and additive manufacturing, have significantly reduced the costs associated with producing composite materials. These advancements enhance efficiency in the manufacturing process, making it easier to produce complex composite structures at scale. As a result, composites are becoming increasingly viable for large-scale applications in the aerospace industry. The ability to produce high-quality composites more economically allows manufacturers to leverage their unique properties—such as lightweight strength and corrosion resistance—across a wider range of aerospace components. This shift not only supports the development of innovative aircraft designs but also contributes to improved performance and sustainability in aviation. Overall, the evolution of manufacturing methods is transforming the role of composites in aerospace engineering, paving the way for their broader adoption and application [18, 24, 37, 38].

6.1.1.8 Sustainability

Composite materials play a vital role in sustainability initiatives by reducing fuel consumption and lowering CO_2 emissions in the aerospace industry. Their lightweight properties contribute to improved fuel efficiency, which is essential for minimizing the environmental footprint of aircraft operations. Furthermore, there is ongoing research aimed at developing recyclable composites, which would further mitigate their environmental impact. By creating materials that can be reused or recycled at the end of their lifecycle, the aerospace sector can enhance its commitment to sustainability. These advancements not only support regulatory compliance but also align with the industry's broader goals of reducing greenhouse gas emissions and promoting eco-friendly practices in aviation. Overall, composites are integral to fostering a more sustainable future in aerospace engineering [28, 34, 39, 40].

6.1.2 Overview of Smart Functional Composite Polymer Materials

Smart functional composite polymer materials are innovative materials that integrate polymers with smart fillers or additives, enabling them to respond dynamically to environmental stimuli. Their numerous uses in a variety of industries, such as aerospace, biomedical engineering, and environmental sustainability, have drawn a lot of attention to these materials. This overview highlights the key characteristics, manufacturing techniques, applications, and future prospects of smart functional composite polymer materials [41].

6.1.2.1 Key Characteristics

Smart functional composites possess the remarkable capability to respond to external stimuli such as temperature, moisture, light, or mechanical stress, enabling dynamic and tailored functionality. These responses manifest as alterations in shape, mechanical properties, or even color, depending on the specific stimulus and material composition. By integrating smart fillers like piezoelectric fibers, shape memory alloys, or conductive nanoparticles into a polymer matrix, their functional range is significantly enhanced [20, 42].

For example, piezoelectric fibers embedded in these composites generate electrical signals when subjected to mechanical stress, making them ideal for applications in sensors and actuators. Similarly, shape memory alloys can restore a predefined shape upon exposure to specific temperature ranges, while conductive nanoparticles enhance thermal and electrical properties, expanding the material's use in energy systems and aerospace. These innovative materials bridge the gap between traditional composites and advanced responsive systems, making them pivotal in modern engineering applications [21, 43].

6.1.2.2 Manufacturing Techniques

The production of smart functional composite polymer materials relies on advanced manufacturing techniques to achieve optimal performance and scalability. Key methods include the following [42, 43]:

1. Additive manufacturing (3D printing): This method enables the layer-by-layer construction of intricate geometries, allowing precise control over the distribution and properties of materials. It is particularly effective for creating lightweight and multifunctional structures.
2. Electrospinning: Electrospinning produces ultrafine nanofibers, enhancing the mechanical strength, porosity, and surface area of the composites. These fibers are particularly valuable for applications requiring high sensitivity or filtration efficiency.
3. Melt extrusion: Widely used for thermoplastic composites, this technique ensures the even distribution of intelligent fillers, such nanoparticles, in the polymer matrix, resulting in enhanced electrical or mechanical properties.
4. Solution mixing: This approach involves breaking down the polymer and additives in a solvent to create a homogenous blend. Upon solidification, the smart components are uniformly embedded in the matrix, improving the material's functional consistency.

6.1.2.3 *Applications*

Smart functional composite polymer materials are increasingly vital across multiple industries due to their advanced and versatile properties, namely [20, 21, 44, 45]:

1. Biomedical applications: In healthcare, these composites serve as the delivery of drugs systems, scaffolds for tissue engineering, and self-healing materials. For example, hydrogels with self-repairing properties improve tissue scaffold performance by enabling them to heal after damage, ensuring prolonged functionality and reduced maintenance.
2. Aerospace engineering: These materials are crucial in aerospace for their lightweight nature and ability to sense and adapt to environmental changes. Applications include structural components like wings and fuselages, where their multifunctionality aids in monitoring and maintaining structural integrity.
3. Environmental sustainability: Smart composites contribute to eco-friendly advancements by enabling recyclable products and adaptive solutions. For instance, corrosion-resistant smart coatings combine protection with sustainability by being reusable or biodegradable.
4. Energy storage: Conductive polymer composites enhance electronic devices for storing energy, such batteries and supercapacitors, by improving electrical conductivity and energy density, contributing to more efficient and compact energy systems.

6.2 Types of Smart Functional Composite Polymer Materials

Smart functional composite polymer materials are advanced materials that combine polymers with smart fillers or additives, enabling them to respond dynamically to environmental stimuli. These materials exhibit unique properties that allow them to change their behavior or characteristics in response to various external factors. Here is a brief discussion on the types of smart functional composite polymer materials [46–50].

6.2.1 Polymer Matrix Composites (PMCs)

Reinforcing fibers or particles are combined with a polymer matrix to create a specific type of composite materials called polymer matrix composites (PMCs). These materials are engineered to enhance performance characteristics for various applications across multiple industries. The polymer serves as the matrix phase, while the reinforcement typically consists of small fiber particles, which can include materials such as glass, carbon, and Kevlar [50, 51]. The most commonly used polymers in PMCs are thermosetting resins, which provide structural integrity and durability. The matrix material plays several crucial roles: it provides the bulk form of the product, holds the reinforcing fibers in place, and distributes loads uniformly throughout the composite structure. This load distribution is vital for ensuring that the composite can withstand external forces without failure. Various reinforcement materials, such as carbon and boron fibers, are used in advanced PMCs; epoxy resins are frequently used as the matrix. Composites produced by this combination have outstanding mechanical qualities, including high ratios of strength to weight, resilience to fatigue, and resistance to corrosion. Consequently, PMCs are commonly used in sectors where durability and performance are crucial, such construction, automotive, and aerospace. Their versatility and adaptability make them an essential material in modern engineering applications. Several manufacturing methods are available for processing polymer matrix composites, with the choice of method depending on the type of materials used and the specific application requirements. Reaction molding, filament winding, autoclaving, and hand layup are examples of common methods [52–54].

6.2.2 Carbon Fiber Reinforced Polymers (CFRP)

After carbon fiber was discovered in 1964 at the Royal Aircraft Establishment in Farnborough, UK, composite materials became a major part of aircraft components. However, these composites were not used on military aircraft until the late 1960s, when they were first tested. Early examples of such applications were doors, rudders, spoilers, and trim tabs. As composites use and knowledge grew, advancements in fiber and matrix materials, including thermosets and thermoplastics, enabled carbon fiber-reinforced polymer (CFRP) composites with enhanced mechanical properties. This development has made it possible to employ CFRP composites in primary aircraft structures instead of more traditional materials like aluminum and titanium alloys [55, 56]. Approximately 5 to 6 μm in diameter, high-strength, high-modulus carbon fibers are made up of tiny crystallites of turbostratic graphite, an allotropic type of carbon. Strong covalent bonds (~525 kJ/mol) and weak van der Waals forces (<10 kJ/mol) hold the hexagonal layers that make up the graphite structure together. A notable variation in mechanical properties is caused by this anisotropic structure: the c-axis, which is perpendicular to the basal planes, has a Young's modulus of just 30 GPa, but the in-plane Young's modulus along the α-axis is approximately 1,000 GPa [57–60].

6.2.3 Glass Fiber Reinforced Polymers (GFRP)

Glass, carbon, aramid, basalt, paper, wood, asbestos, and other fibers have all been developed or identified over time to construct complex fiber-reinforced polymer (FRP) composites. The most commonly used of these in the production of FRP are glass fibers (GFs) and carbon fibers (CFs). Glass fibers perform exceptionally well under high tensile stress but are less effective under compression due to their brittle nature. Conversely, plastic materials lack tensile strength but exhibit excellent compression resistance. By combining these materials, glass fiber-reinforced polymers (GFRPs) are formed, offering the ability to handle both tensile and compressive loads effectively [61, 62]. Because of their improved performance, GFRP composites are becoming more and more common in a variety of applications, such as sheet molding compounds, automotive parts, sports equipment [63], boat and ship constructions, aerospace components, and electrical [64], sound, and thermal insulation. Figure 6.2 shows the various glass fiber classifications and physical attributes.

The glass fibers provide strength, stiffness, and durability, whereas the fibers are bound together and shielded from harm by the polymer matrix. GFRPs are renowned for its adaptability, lightweight design, and mechanical qualities, which make them ideal for use in various industries. They come in different types of glass fibers, including E-glass for electrical insulation, S-glass for strength, and AR-glass for cement applications, each tailored for specific purposes [65–68]. The primary advantages of GFRPs include strong resilience to fatigue and tensile strength, and impact resistance, as well as excellent thermal insulation and electrical properties. They are also corrosion-resistant and perform well in harsh environments. Additionally, GFRPs are lightweight, offering significant weight savings compared to metals like steel and aluminum. The materials can be molded into complex shapes, providing both aesthetic and functional versatility, with applications spanning the industries of renewable energy, maritime, automotive, and aerospace. Several production methods, including spray-up, manual lay-up, and pultrusion, are used to create GFRP products, allowing for

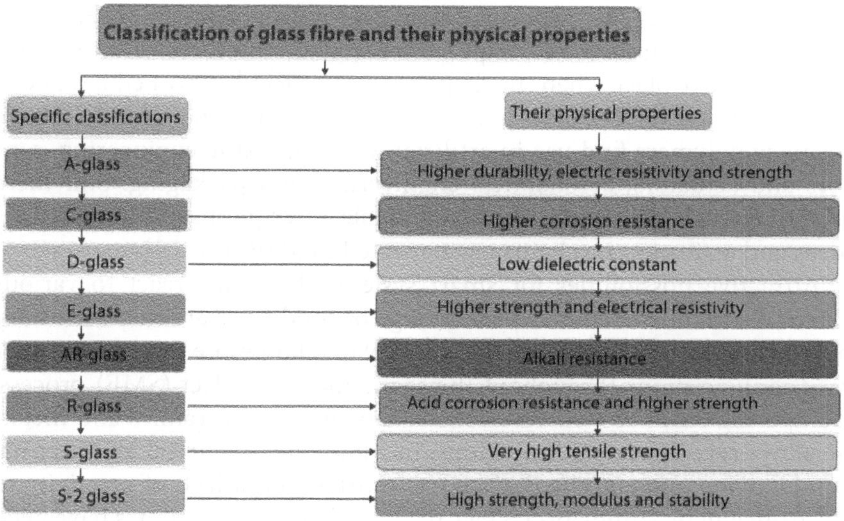

Figure 6.2 Types and physical characteristics of different glass fibers.

flexibility in production [69]. Despite their numerous benefits, GFRPs face challenges such as brittleness, environmental concerns regarding disposal and recycling, and relatively high manufacturing costs. Additionally, their thermal resistance is limited, and moisture absorption can compromise the integrity of the material. However, ongoing research [70] aims to address these issues, including the development of recyclable matrices, the use of hybrid composites, and innovations in advanced manufacturing methods like 3D printing. With continuous advancements, GFRPs are expected to play a pivotal role in the future of engineering, opening new possibilities in emerging fields.

6.2.4 Shape Memory Polymers (SMPs)

Shape memory polymers (SMPs) are advanced materials that can be momentarily bent and bonded into a shape, but they can revert to their original shape when the external stimulus is removed [71]. Their remarkable properties, such as their structural versatility, lightweight nature, cost-effectiveness, ease of processing, high elastic strain (exceeding 200%), biocompatibility, and biodegradability, make them highly desirable for use in a range of industries, including aerospace, industry, textiles, and medicine. SMPs can react to a range of external stimuli, including heat, light, electricity, magnetic fields, chemical changes (such pH variations), and humidity [72–74]. Aspects of shape memory polymers (SMPs), such as their activation methods, required molecular structures, and the alterations that take place when the shape memory effect is activated have been the subject of several reviews in recent years. As shown in Figure 6.3, shape memory polymers find use in artificial muscles, textile engineering, sensors and actuators, aeronautical engineering, and packaging. Still there are not enough concise and understandable descriptions of the structural ideas and activation mechanisms connected to SMPs [71, 75].

Attractive possibilities for smart systems that must react to various external stimuli and control over a range of conditions are shape memory polymers (SMPs), which have the capacity to retain more than two temporary configurations throughout the shape memory effect (SME) process. Comparing SMPs to well-researched shape memory ceramics and metallic alloys reveals several limitations, including low electrical and thermal conductivity, poor mechanical characteristics, and a lack of reactivity to electromagnetic stimulation. Over the past two decades, efforts have been focused on creating new polymers with improved SME for improved performance. This includes changing the structures of polymers using various

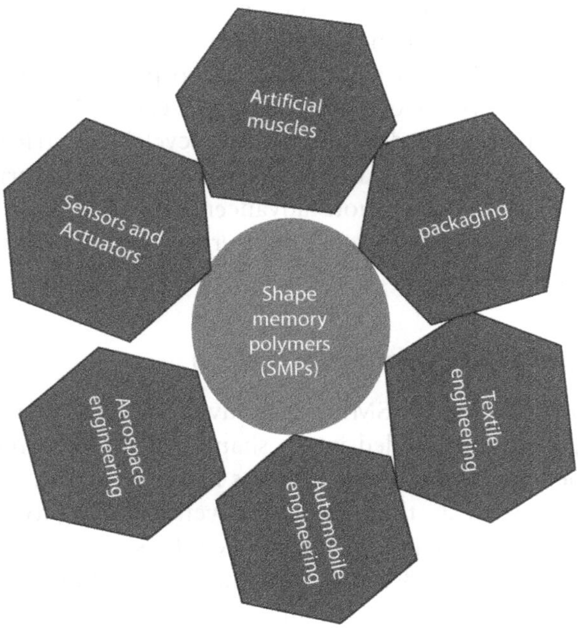

Figure 6.3 Applications of shape memory polymers (SMPs).

processing methods and adding various fillers to increase mechanical strength, electrical conductivity, and thermal conductivity [76–79].

It is currently possible to trigger the shape memory effect (SME) in shape memory polymers (SMPs) using three distinct external stimuli: light (photo-responsive and photo-thermal responsive SMPs), heat (thermo-responsive SMPs), and chemicals (chemo-responsive SMPs). The textile, aerospace, and medical sectors can all benefit from these [80].

6.2.5 Self-Healing Polymers (SHPs)

Due to their unique dynamic properties, stimuli-responsive materials have the potential to significantly contribute to the development of future technologies [81]. They are able to change characteristics including color, size, shape, conductivity, and permeability because of their ability to sense and react to internal or external stimuli, including biological surroundings, temperature changes, pH shifts, mechanical stress, and electric or magnetic fields. These adaptable properties provide creative answers to problems that conventional materials are unable to solve. The development of self-healing polymers (SHPs), which draw inspiration from natural systems, has accelerated due to both scientific curiosity and the rising need for long-lasting, sustainable materials. To meet this need, future advancements will require not only meticulously designed and engineered self-repairing materials but also everyday polymers capable of healing autonomously under normal usage conditions [81–83].

Recent self-healing strategies often rely on the physical or chemical incorporation of repair mechanisms. These approaches generally fall into four categories: (1) embedding capsules filled with reactive liquids that release healing agents upon damage, repairing the material [84–87], (2) using dynamic, reversible chemical bonds that reform after being disrupted [88–90], (3) dispersing nanoparticles, such as superparamagnetic materials, that respond to external stimuli like electromagnetic radiation, generating heat to enable flow and diffusion [91], and (4) achieving self-repair driven solely through interchain van der Waals interactions, which in certain compositions without additional components create key-and-lock junctions [92].

Self-healing can be initiated through various triggers, including heat, light, pH shifts, redox reactions, mechanical compression, or moisture [88–92]. Notably, some materials demonstrate the remarkable ability to self-repair autonomously under ambient conditions, eliminating the need for external intervention. These advancements highlight the versatility and potential of SHPs in addressing future material challenges [81, 84, 87].

Self-healing polymers (SHPs) possess the intrinsic capability to recover their properties after being subjected to mechanical, thermal, or other forms of damage, thereby extending their lifespan and functionality [81, 93, 94]. The shape memory effect (SME) is frequently used by polymers intended to fix dents or scratches in order to speed up the healing process [89]. Notably, deformations that seem permanent when caused below the reversible transition temperature can be repeatedly recovered thanks to reversible plasticity shape memory (RPSM) [81]. When a material exhibits sufficient fracture toughness or resistance to crack formation upon damage, deformation can be reversed through SME activation. This is achieved by utilizing the conformational entropic energy stored during the damage event to bring damaged surfaces into close proximity [81]. Once this proximity is established, subsequent chemical or physical self-healing mechanisms restore mechanical strength.

6.2.6 Conductive Polymer Composites (CPCs)

Combining conductive fillers, for example, carbon nanotubes (CNTs), carbon fibers (CF), carbon black (CB), or suitable particles that provide efficient conductive pathways inside the polymer matrix with an insulating polymer matrix results in conductive polymer composites (CPCs) [50, 95]. Numerous advantageous qualities are provided by these composites, such as excellent mechanical performance, corrosion resistance, lightweight structure, and high electrical conductivity [95]. Nonetheless, the shape and configuration of the conductive network inside the composite matrix significantly affect the electrical conductivity of CPCs [96]. Modern advancements have enabled CPCs to find applications in various fields, such as sensors, electronic circuit components, batteries, fuel cell electrodes, and bipolar plates for fuel cells [95–98]. Conductive polymer composites (CPCs) come in various types, each tailored for specific applications. Carbon-based CPCs utilize materials like graphene, carbon fibers, carbon nanotubes (CNTs), or carbon black to create conductive pathways and enhance properties. Metal-based CPCs include metal powders or metal-coated fillers for superior conductivity. Hybrid CPCs combine different fillers, such as CNTs with graphene or carbon black with metal powders, to optimize performance. Ionic conductive CPCs incorporate ionic fillers or electrolytes for applications like sensors or batteries. Conductive polymer blend composites mix inherently conductive polymers (e.g., polyaniline) with insulating matrices, while nanocomposites utilize nanoscale fillers for improved conductivity and structure. Finally, intrinsic conductive polymer composites directly integrate conductive polymers into the matrix, eliminating the need for additional fillers [99, 100].

6.3 Properties and Characteristics

6.3.1 Mechanical Properties

This section major around the mechanical properties of structural materials with additional contribution to polymer composite materials in order to improve their importance with respect to the conventional role of structural materials. It is common that the composites (hybrid) properties are largely related to the associated properties. However, there are many submissions in researches, simulated experimentally, that the hybrid properties are not applicable to the mixing rule. This shows beyond constituent's properties, there are other metrics negating the end physicochemical hybrid properties. The host and fillers' interfacial interactions are one that largely affect composite properties. In this study, we summarize the types of interactions between the linked components, how to improve them, how to assess them, and how interfacial interactions affect the properties of electrical, mechanical, thermal, and composite materials. Polymers are molecules developed from repeated units long chains designated as monomers. Their flexibility, low weight and minimal cost of production, make them suitable for usage in day to day activities such as painting, automobile industries, food packaging, etc. Although the monomer structure and polymerization selection methods grants quality control on polymer properties section, some functions are not achievable by polymer itself. Hence, polymers achieve new properties when compounded with other additives. This completed product is known as composite [101].

One thing they all have in common is the employment of fillers, which can change the properties of a matrix. Micrometer-sized inorganic fillers including glass beads, calcium carbonate, and talc have been widely used in common composites to increase mechanical properties [102–104]. The aforementioned properties can be extensively made better by reducing size of fillers to a minimal ratio [102, 105]. Composite mechanical properties are largely link to filler-matrix, interfacial adhesion, size of particle, and loading [106–108]. For instance, there is high dependence of both toughness and strength on the adhesion interface. Consequently, the dispersion, geometric dimensions, interfacial adhesion, and others, are vital in enriching the mechanical properties [109–111]. A review can carry out on the mechanical properties using conventional methods like dynamic mechanical analysis commonly called DMA [112, 113]; other properties like shear tests, tensile, and compression [114–116]; copper grid technique which are alternative technique [117, 118] and elastic buckling that are strain-induced create instability for mechanical metrics [119–121]. For hybrid polymers, stiffness does not largely affect interfacial bonding degree [122, 123]; it is not reviewed here. One major leading influential part of mechanical properties of any composite is the strength and strength cannot be overemphasized.

6.3.1.1 *Strength*

The composite tensile strength rely on efficiency transfer of the stress between the composite geometry. Strength improvement depends on the load movement toward the fillers [124, 125]. Smaller particles have larger contact areas and consistent filler volume percentages, which leads to transfer regions with high stress [106]. Additionally, the interfacial bonding strength between the composite ingredients affects the load transmission efficiency [106, 125]. The strength of composites with weakly bound fillers will be lower than that of composites with significant interfacial contact [122]. This is because de-bonding at the interface creates discontinuity, which makes it difficult for the filler to effectively support the imposed load. Numerous studies have shown that altering the filler surface increases dispersion and interfacial interaction, which in turn raises the composites' tensile strength [126–130]. These studies also suggest that adding chemical bonding to filler–matrix interfaces can significantly increase the strength of the composites [130, 131].

6.3.1.2 *Stiffness*

Materials ability to return to its original shape after compression or tension by an external force is referred to as stiffness. It also means the material's ability to resist this outer forces and still return to its origin. The said forces are stretching, bending, and other various forms of strain. Man-made structures mechanical metamaterials are with counterintuitive mechanical properties that begins in the geometry of their unit cell in place of the properties of individual component. The typical mechanical metamaterials are commonly linked with the four elastic constants, shear modulus G, the Young's modulus E, Poisson's ratio t, and bulk modulus K, with all these largely liked with the stiffness, rigidity, and compressibility of a material from an engineering perspective.

6.3.1.3 Toughness

There is a submission on nanofillers role in growth of tough polymeric [133, 134]. However, the enhancement of the composites strength are accompanied with sacrificing products toughness [135]. Similarly, toughening agents like rubbers employed to improve the extensibility and the fracture polymers resistance, reduce the product strength [136, 137]. Hence, the equilibrium for these two properties should be considered when composites properties are being designed [138, 139]. Fracture behavior of polymer nanocomposites rely on the various polymeric matrix [106], fillers shape and size [132], and the interactions of the interfacial [132]. For example, the enhancement of the interfacial adhesion largely enhances the toughness between the thermoplastic matrix and filler, but not in composites with thermosetting matrix [106]. Still the glassy polymeric matrix is unclear during the toughness and strength simultaneous enhancement [132].

6.3.2 Thermal Properties

Composites thermal properties can differ with thermomechanical service due to changes to the interface between reinforcement and matrix. Further research is carried out into thermal property modeling properties of hybrid materials of various forms; however, the match to demonstrate metrics is often not suitable. It is related to the belief on model creation, frequently seen as simplistic. Technically, the interfaces state is unusual or may differ with the employed conditions, for example thermal or mechanical rotation. Examination of the available model is the purpose of the present review and the projections behind resulting in the sensitivity of the thermal properties to known structural changes.

6.3.2.1 Thermal Conductivity

Several literatures verified that the composite interphase play an important role in their thermal conductivity [140–146]. When two distinct materials come into contact, heat transfer often occurs when there is a temperature discontinuity [147]. This great stride was initially viewed at the interface of a helium of metal and liquid [148], subsequently known to be the interface of two solids [149]. Therefore, the interfacial area overwhelmingly cause effect on the composite thermal conductivity [150, 151]. Therefore, whatever affects the interfacial space (e.g., particles geometry [152–158], aggregation [159–161], interfacial pressure [162], roughness [163–165], and the interactions strength at the interfaces [166–173] in the composites would create a significant change on their thermal conductivity. Table 6.1 in this work shows the data negating the interfacial interactions, the resulting impact on the hybrid thermal conductivity.

Table 6.1 Filler types, functionalization improves polymer conductivity.

Filler	Matrix	TC change	Fraction	Reference
(BN)[a]-coated polymethylsilsesquioxane	Epoxy	+ 9 times	30 vol%	[174]
MWCNTs@ SiO_2-g-BMI[b]	Epoxy	+ 125%	1.25 wt%	[175]
Noncovalent functionalization with pyrene carboxylic acid, BNNTs[c]–BNNSs[d]	Epoxy	+ 95%	1 vol.%	[176]
Silica-coated aluminum nitride	Epoxy	+ 10 times	50 vol%	[177]
Noncovalent functionalized graphene nanosheets	Epoxy	+ 267%	1 phr	[178]
GA-II[e] anchored graphene (noncovalent)	Epoxy	+ 12 times	2 wt%	[177]
GNP-C750	Epoxy	+ 115%	5 wt%	[178]
Alumina-decorated graphene nanoplatelets	Epoxy	+ 677%	12 wt%	[179]

(Continued)

Table 6.1 Filler types, functionalization improves polymer conductivity. (*Continued*)

Filler	Matrix	TC change	Fraction	Reference
Noncovalent functionalized graphene flakes	Epoxy	+ 665%	10 wt%	[180]
Alumina-coated graphene sheet	PVDF	+ 192%	40 wt%	[180]
Functionalized GNPs	Polyphenylene sulfide (PPS)	+ 19 times	29.3 vol%	[181]
Graphene-graphene oxide	Polyamide-6	+ 6 times	10 wt%	[182]
SiCNWs[f]-graphene sheets	PI	+ 138%	7 wt%	[183]
ApPOSS–graphene[g]	Epoxy	+ 37.60%	0.25 wt%	[184]
TCA-rGO[h]	Polyamide	+ 19 times	5 wt%	[185]
Noncovalently modified graphene nanosheets	PU	+ 34%	0.608 wt%	[186]

[a]Boron nitride.
[b]1,10-(Methylene di-4,1-phenelene) bismaleimide (BMI) was grafted to core-shell MWCNT@SiO_2-NH_2.
[c]Boron nitride nanotubes.
[d]Boron nitride nanosheets.
[e]Bio-based epoxy monomer (GA-II) synthesized from renewable gallic acid.
[f]One-dimensional silicon carbide nanowires.
[g]Aminopropylisobutyl polyhedral oligomeric silsesquioxane covalently grafted graphene.
[h]Modified reduced graphene (rGO) with titanate coupling agent.

6.3.2.2 *Thermal Stability*

What creates suitability for high temperature applications is thermal stability and better flammability performance. Furthermore, knowing the collapsing temperatures is essential for a better process design because many polymer hybrids are created at high temperatures by melt mixing [106, 174]. The introduction of reinforcing compounds improves the matrix's thermal stability in a number of ways. They are required for flame retardation applications because, among other things, they can function as a barrier [175–177]. Second, they have the ability to create a network that protects the polymer from deterioration [178–182]. Third, they can act as radical traps [183]; fourth, they can alter the microstructure of the product [106, 183]. Furthermore, high adhesion between the matrix and filler reduces the mobility of polymer chains, which in turn lowers the rate of breakdown [184]. Hence, employing any means which can increase the interaction strength between the hybrid constituents help make thermal stability to improve.

6.3.3 Electrical Properties

The formation of continuous conductive pathways is a result of well-dispersed fillers and directly enhance electrical conductivity (EC). Inversely, the host interaction with a working filler creates an insulating layer on filler's surface, which has an effect on EC enhancement.

6.3.3.1 Conductivity

Unlike the mechanical and thermal composites properties, that presents unilateral or distribution that are homogenous with adhesion that are strong between the matrix and the fillers is needed, electrical conductivity (EC) is centered on continuous electrically conductive network formation between the fillers [185]. The electrical conductivity enhancement shows in three major phases, () prior to, (ii) within, and (iii) after percolation threshold [186]. Step by step aggregation and link through CFs, the EC improves a bit through tunneling effect. Though there is still no total direction for conduction. The first conductive path (red lines in state c) is produced by further increasing the filler mount. This level's filler fraction volume is referred to as the percolation threshold. More conductive pathways are created when filler is added to the host, which improves the conductive network. The size, aspect ratio, dispersion, alignment, and interfacial interactions of the fillers all affect the percolation threshold, which is dependent on a sudden drop in electrical resistance [187–189].

6.3.3.2 Resistivity

The experimental resistivity of composites increased with increasing volume proportion and decreasing reinforcement size, as predicted theoretically. The composite resistivity increased as a result of the deformation zones surrounding the reinforcement that contained residual stresses and dislocations. By combining thin slices of composite cells, theoretical models of electrical resistivity prediction were created for matrix composites of metal reinforced with continuous fibers, short fibers, and particulates.

6.3.4 Multifunctional Properties

Composite materials are continuously employed for applications of high-end like space ships, aviation technology, and manufacturing equipment that are heavy. The deployment of materials originated from composite is understudied in recent times in multifunctional composite materials commonly refer to as MFCMs field. There is consistent progressive developments, innovations, improvements, despite extensive destructive and non-destructive testing, demonstrating the materials' resilience and long-term endurance. and replacement of metals.

6.3.4.1 Self-Sensing

One of the materials that can detect their own state is self-sensing composite materials, such as stress, strain, and damage, and convert them into properties of electrical output. This helps them to monitor their health and point potential mechanisms that can deteriorate. For the purpose of health monitoring structures of civil engineering, self-sensing composites are a potential technology. They can react to outward factors like temperature variations, loading conditions, and environmental variations. Piezoresistivity principle is where self-sensing composites work by detecting changes in electrical resistivity. Summarily, self-sensing is the process of incorporating conductive fillers into the matrix, using an electric field, and connecting the composite with analogous circuit models.

6.3.4.2 Self-Healing

The interaction between epoxy resin and POSS (polyhedral oligomeric silsesquioxane) nanocages determines the component used to trigger self-healing processes ingrained in multifunctional materials thermosetting: reversible hydrogen bonding. The integrated MWCNT (multiwalled carbon nanotubes) phase composition of the multifunctional resin can be advantageously utilized to support approximately 400% of the healing efficiency. MWCNT in the matrix is the cause of the increased mobility of chains from finely interpenetrated domains where enhanced healing efficiency is provided by the reversible hydrogen bonds.

6.4 Aerospace Applications

Intelligent materials also known as smart materials are gaining advantage continuously in several industries not excluding aerospace [190]. These emerging materials like memory capabilities, self-sensing, manifold functions, and self-sensing is what makes the unique features. For a while, there is no reconsideration of these intelligent materials. Hence, it is a noteworthy study to consider aerospace application of these materials especially for structural components [190].

6.4.1 Structural Components

The components that make up an airplane's structure are designed to transfer and withstand applied loads, provide for an aerodynamic form, and shield systems, passengers, payload, and other components from environmental conditions encountered during flight. These requirements result in thin-shell constructions in many aircraft, where the exterior surface or shell skin is typically supported by longitudinal stiffening and transverse frames to help it withstand compressive, torsional, and bending stresses and prevent buckling [191]. These kinds of constructions are called semimonocoque, whereas thin shells that can only withstand weights by using their skins are called monocoque. Three groups of structural elements can be distinguished in an aircraft: the fuselage, the wings, and the control surfaces. Through mechanical joints and aerodynamic linkage, these three sections are interlaced.

6.4.1.1 Wings

The wings use the lift generated by their airfoil shape to support the plane. These are importantly, cantilever beams undergoing bending and torsion [191]. Depending on the plane's size, one or more I-beams may run from the fuselage to the wing tip with the necessary taper to fit the shape of the airfoil. For small planes, metal tubes are used instead of I-beams. Additionally, the top and bottom surfaces include longerons that run parallel to the beams. The ribs (diaphragms) keep the airfoil in its proper shape and separate the wing's length into compartments. Aerodynamic loading producing shear force is what the wing is subjected to [191].

6.4.1.2 Fuselage

One characteristic of fuselages is their circular or oval cross-sections, which vary in size as the fuselage extends [191]. In order to preserve the oval shape and provide support points for the wings, bulkheads are a series of circular (or oval) metal rings arranged parallel to one another.

6.4.1.3 Control Surfaces

The sections of an airplane the pilot uses to operate is called the control surfaces, i.e., operations like to accelerate, decelerate, to taxi, aviate, bank, and land. By forcing differences in air pressure, these various sections of the aircraft uses air pressure (the air surrounding) it to take whatever action the pilot wishes [191]. What makes flight possible is the ability to manipulate the various sections of the aircraft.

6.4.2 Functional Components

There is contribution of components of all aircraft to the functionality of the aircraft [191]. The aircraft dynamic parts such as aileron, Wings, flaps, spoiler, fuselage, jet engine, rudder, elevator, cockpit, horizontal stabilizer, vertical stabilizer, etc., are the operational parts which complement the whole aircraft system [192].

6.4.2.1 Actuators

One of the major sources of worry for actuators that are electromechanical is usually deployed in the aerospace application demanding conditions. Some of the major effects are environmental factors, excessive loads, manufacturing defects and lubrication issues [191, 193]. Since all of the actuator components are affected and currently have the fewest effective diagnostic and prognostic techniques developed; this class of problems has long been the main focus of study.

6.4.2.2 Sensors

An extremely wide and complicated sensor spectrum on a modern aircraft, either military or civil is known to be extensively wide and complicated. The emerging innovations and technology developments, such as fly-by-wire, places new requirements on reliability and performance of sensors. Advance aerospace system is needed for some of the modern technology that are now embedded to ensure quality assurance of the sensors. Future developments in sensors will be overwhelmed by certain required aerospace system, such as upgraded functionality of the system performance, optimum reliability, maintainability, and integrity [194].

6.4.2.3 Energy

There is a growing challenge for the aerospace industry to navigate into new alternatives that are sustainable for a greener and reliable future. This chapter seeks to present an overview of the different ways sustainable energy is being incorporated in the aerospace community. The different form of energies included here are electrohydrodynamic, biofuels, hybrid, solar, electric, hydrogen, and nuclear [94]. Not only is the aerospace industry including alternative greener ways of operation, but also giving back by making better the recent energy generation techniques. The use of materials and techniques for sustainable manufacturing is other ways to inculcate sustainability.

6.4.2.4 Storage

The leading donor in the growth and implementation of NASAs energy storage technologies for research programs is the NASA Glenn Research Center (GRC). NASA GRC has assisted efforts in technology for the fuel cells and battery improvement. The extraordinary role being played in the growth and implementation of technologies for energy storage at the NASA GRC electrochemistry branch, in a joint-role with other centers associated with NASA, academia, government agencies, and industry. This chapter underscores the need for fuel cell technologies and batteries at the NASA Glenn center of research. A number of systems required are being covered to ascertain that NASAs requirement for a broader variety of systems are achieved. A broad knowledge covered are lithium-based batteries, including PEM—proton exchange membrane, fuel cells, and various activities related to nanotechnology.

6.4.3 Thermal Management Systems

Thermal management refers to the technologies and instruments used to keep a system within its operating temperature range. Electronic equipment are necessary to prevent overheating, and thermal management usually dissipates surplus heat. Majority of electronic systems dissipate heat, and components that are internally sensitive can be spoilt with excessive thermal energy accumulating the device. Similarly, electronics can be affected as a result of external interference.

6.4.4 Self-Healing Coatings and Materials

A significant increase in the creation of self-healing composites has been noted, and numerous self-healing materials have been examined and explored since the 2000s [195]. Self-healing nanocomposites and polymers are a significant class of responsive materials. One of these materials' properties is their ability to repair their damage in a reversible manner. In aerospace applications, thermosets and thermoplastic polymers have been reinforced with nanocarbon nanoparticles to enable self-healing of structural damage.

6.5 Manufacturing Techniques

Due to their exceptional ability to balance high strength and low weight, composite materials are frequently utilized in aerospace applications [196]. Common manufacturing methods for these composites include hand layup, autoclaving, resin transfer molding, and automated layup. Although autoclave processes are favored in the aerospace sector due to their high precision, there is an increasing shift towards out-of-autoclave techniques [197]. While these conventional methods remain dominant due to their simplicity and cost-effectiveness, advanced techniques like additive manufacturing (AM) has also emerged, enabling the creation of customized composites with enhanced functionality, integrating diverse materials and emerging AI technologies for various applications with minimal waste [198]. The industry is also exploring bio-based composites to address environmental concerns [199]. Despite advancements in automation, many composite manufacturing processes still lack consistent high levels of automation throughout their chains [200]. The future of composite manufacturing in aerospace is shifting towards greater automation and smart technologies, with an emphasis on incorporating Industry 4.0 advancements such as big data and machine learning [197, 200]. This transition seeks to boost efficiency and productivity and lower costs in the manufacturing of composites for aerospace applications. This section highlights some of the common techniques currently used in the industry.

6.5.1 Conventional Methods

6.5.1.1 Hand Layup

This is an easy and affordable way to for manufacturing composites, involving manual layering of fibers and resin (Figure 6.4), but it offers lower mechanical properties and surface finish compared to advanced techniques [201]. Vacuum bagging offers improved mechanical and fire resistance properties over hand layup, making it ideal for automotive and railway applications [202]. Although newer automated techniques provide higher precision and scalability, hand layup remains vital for prototyping, specialized production, and applications requiring design flexibility [203]. The Hand Layup Process, resin poured on the fiber placed onto the mold and roller smoothen the resin for even distribution on the fiber is illustrated in Figure 6.4 which corresponds with the works of some researchers [204, 205].

6.5.1.2 Resin Transfer Molding (RTM)

This is a flexible manufacturing method of fiber reinforced composites, offering a more controlled process compared to hand layup. A recent advancement, RTMsim, a Julia-based simulation tool was introduced for optimizing the RTM filling process [206]. To address recycling challenges, thermoplastic resins are being investigated as substitutes for thermosets in resin transfer molding (RTM). These resins use low-viscosity formulations made from cyclic esters, methyl methacrylate, and cyclic oligomers [207]. A 3D woven thermoset composites manufactured *via* RTM shows

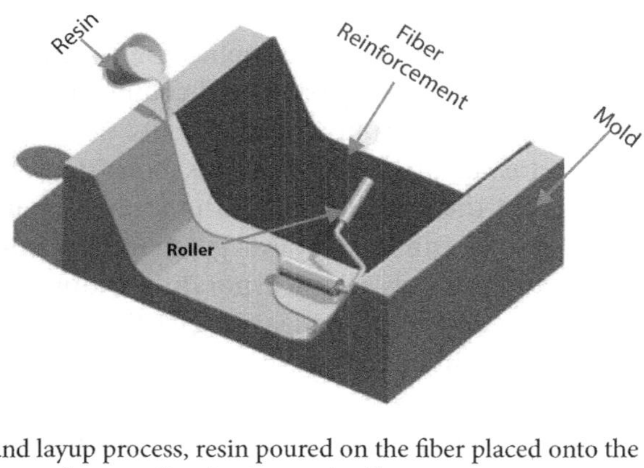

Figure 6.4 Hand layup process, resin poured on the fiber placed onto the mold and roller smoothen the resin for even distribution on the fiber.

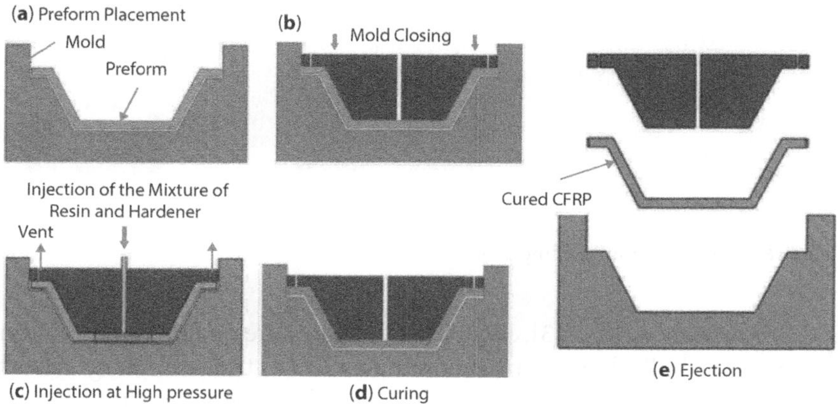

Figure 6.5 Resin transfer molding process.

potential when multi-scale analysis was conducted, with five key areas identified for future research, including developing polymer constitutive models, verifying simulation accuracy, and improving computational efficiency [207]. Figure 6.5 shows the resin transfer molding process: (a) the preform is placed into the mold, taking its shape (b), the upper mold is positioned to close and compress the preform (c), a high-pressure mixture of resin and hardener is injected to saturate the preform, (d) the wetted preform is cured inside the mold, and (e) finally, the mold is opened, and the cured CFRP component is removed as also supported by some researchers [208, 209].

6.5.1.3 Compression Molding

Compression molding is a versatile manufacturing method for producing composite materials and structures. It enables the use of high-toughness elastomers and allows for faster production cycles compared to ambient casting methods [210]. The process is particularly challenging for materials with high fiber content due to complex rheology, temperature gradients, and fiber orientation effects [211]. When producing hybrid composites using the compression molding technique, a numerical model can be employed to analyze the thermomechanical and rheological behaviors of both continuous and discontinuous fiber-reinforced layers [212]. The technique has also been applied to fabricate highly filled wood plastic composites, where processing parameters such as compression pressure and mold temperature significantly influence mechanical properties, dimensional stability,

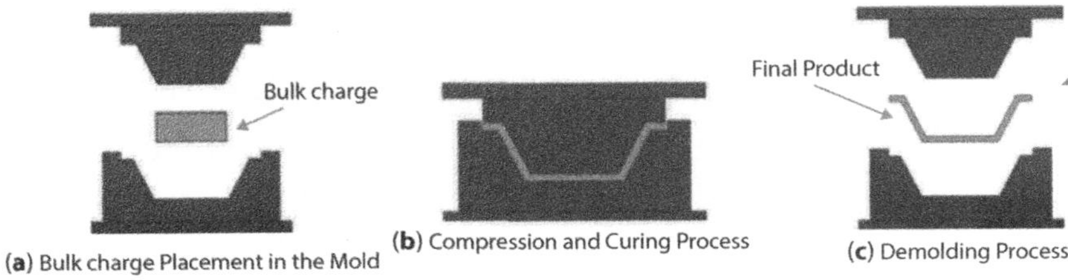

Figure 6.6 Compression molding process.

and production yield [213]. Figure 6.6 indicates the compression molding process: (a) a bulk charge is placed in the mold, (b) the bulk charge is compressed by the upper mold, taking the mold's shape while curing under pressure, and (c) after curing, the upper mold is removed, and the finished product is extracted as illustrated by some researchers [214, 215].

6.5.2 Advanced Methods

6.5.2.1 Additive Manufacturing (AM)

Among the most transformative manufacturing techniques, additive manufacturing (AM) have stretched the potential for creating smart functional composite polymer materials for aerospace applications with high strength, and long-term stability. Aerospace structures can be manufactured using fiber-reinforced auxetic composites with deployable characteristics, such as time-dependent structural adaptations, which can be fabricated through additive manufacturing of functional polymer nanocomposites using 4D printing technology [216]. This technique enables the creation of geometrically intricate structures that would be either impossible or extremely costly to produce using traditional methods. Polymer nanocomposites are being developed for various additive manufacturing (AM) techniques, such as fused filament fabrication and selective laser sintering, which support the production of multifunctional parts with complex designs [217]. AM enables design flexibility, reduces waste, and facilitates manufacturing of complex customized aerospace components with polymer composite materials [218]. Shape memory polymer composites (SMPCs) became ideal in aerospace engineering applications due to their outstanding characteristics, such as significant recoverable deformation and enhanced mechanical properties, [219]. The integration of functional materials and nanostructures in SMPCs allows for controllable remote actuation and improved performance [219]. Additionally, [220] discover a 3D printing of continuous carbon fiber reinforced polymer composites, highlighting protective and intelligent designs for aircraft skins.

6.5.2.2 Electrospinning

A versatile method for creating ultrafine fibers, electrospinning finds usage in a number of industries, including wound dressings, tissue engineering, and medication administration [221, 222]. A polymer solution and a powerful electric field are used in the technique to create nanofibers with enormous surface areas and modifiable characteristics [222]. Electrospinning has shown potential in improving the solubility and bioavailability of poorly soluble drugs and encapsulating heat-sensitive biomaterials [223]. While electrospinning offers simplicity and efficiency in continuous nanofiber production, challenges remain in achieving consistent fiber structures, scaling up for industrial applications, and addressing sustainability concerns [224, 225].

6.5.2.3 Automated Fiber Placement (AFP)

This is a sophisticated manufacturing method that automates the process of laying continuous fiber reinforcement onto a mold, commonly used for large-scale composite parts, especially in aerospace applications. While AFP offers numerous benefits, it also faces challenges, including defects such as gaps and overlaps, which can significantly affect the mechanical properties of the final product [226]. Recent researches focuses on adopting machine learning, photonic sensing, and thermo-mechanics modeling to optimize AFP processes [227]. AFP is also being explored as a potential replacement for traditional filament winding in composite pressure vessel production, offering greater design flexibility and optimization possibilities. These improvements aim to enhance manufacturing techniques, improve material performance, and widen the application of AFP in various industries, including hydrogen storage for mobile applications [228]. Figure 6.7 illustrate automated fiber placement process, carbon fibers impregnated with epoxy resin, known as tows, are fed from spools. The tows are heated and deposited onto the mold by a compaction roller, which simultaneously applies pressure to ensure proper bonding and consolidation as corroborated by the works of Han *et al.* [229] and Rashid *et al.* [230].

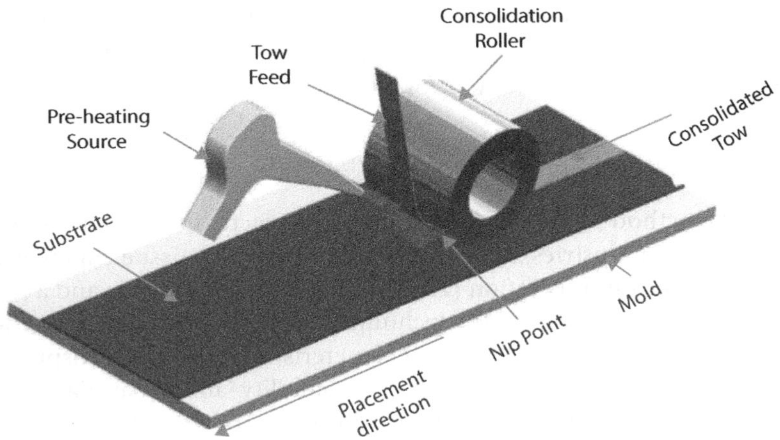

Figure 6.7 Automated fiber placement: carbon fiber deposition and consolidation process.

6.5.3 Nanotechnology-Enabled Manufacturing

Recent advancements in nanotechnology have significantly enhanced the manufacturing and performance of composite materials. Mechanical properties, surface modification, and sensing capabilities can be improved by incorporating nanoparticles into fiber-reinforced polymer composites [230]. Additive manufacturing techniques, particularly vat photo polymerization, enable the production of nanocomposites with complex structures and controlled properties for aerospace, automotive, and medical applications [231]. X-ray CT, infrared thermography, and ultrasonic phased array technologies non-destructive testing methods are crucial elements for quality control and process optimization in Nano-additive manufacturing [232]. A novel approach using programmable spray coating allows for the deposition of patterned nanostructures in composites, enabling tailored interfaces, damage containment, and electrical-thermal conductivity. This method eliminates trade-offs between mechanical and multifunctional properties, offering new possibilities in hierarchical composite manufacturing [233]. Direct laser writing (DLW) is an innovative technique that allows for the creation of nanoscale features on composite materials. DLW can be used to produce nanocomposites with enhanced mechanical and optical properties, such as silica nanoparticle-based materials exhibiting structural color [234].

Innovations like the fabrication of ZnO/CuNi-polyimide composite films demonstrate significant aerospace applications, enhancing mechanical robustness and environmental resistance through integration-focused fabrication techniques [235]. Surface coating involves applying a material layer to composite surfaces to improve properties such as corrosion resistance, wear resistance, and thermal stability [236], self-healing materials, inspired by natural processes, autonomously repair damage using embedded microcapsules or vascular networks that release healing agents when damage occurs [237], while [238] introduced a laser-assisted approach for fabricating polymeric micro-composites with controlled porosity, which is beneficial for lightweight aerospace structures.

6.6 Challenges

Smart functional composite polymer materials offer enormous potential to redefine aerospace technology with multifunctionality and advanced capabilities, however, face several challenges. These consist of corrosion, fretting wear, stress corrosion cracking, and inadequate mechanical qualities [239]. The creation of sophisticated materials is required due to the harsh circumstances found in space, including micrometeorites, fire dangers, space dust, temperature fluctuations, and electromagnetic interference [240]. Furthermore, shielding against electromagnetic interference is a serious issue that necessitates the creation of useful composite materials for particular aeronautical applications [241].

6.6.1 Interfacial Properties and Bonding

Lack of optimization of interfacial properties between the matrix and reinforcement is a great challenge in the development of smart composite materials. In composite materials, interfacial properties between fibers and matrices are crucial, poor interfacial bonding reduce the mechanical performance of the composite [242]. Nano indentation techniques can assess the distribution of mechanical properties and interfacial characteristics in carbon nanotube-reinforced polymers and carbon fiber composites [243].

6.6.2 Scalability and Cost-Effectiveness

While the performance benefits of smart composites are clear, scaling up production to meet the demands of the aerospace industry is a significant challenge. The incorporation of high-performance materials, such as carbon nanotubes and graphene, offers promising applications in aerospace due to their multifunctional capabilities [244] but can significantly increase the cost of production, which could limit the widespread adoption of these materials in aerospace applications. Scalability remains a significant hurdle, with researchers exploring methods like layer-by-layer deposition of 2D materials on glass fiber fabrics at high speeds [245]. The ability to scale up manufacturing without sacrificing material quality will be key to the commercial viability of smart composite materials.

6.6.3 Integration with Existing Aerospace Systems

Integrating smart functional composite materials into existing aerospace systems poses a significant challenge, particularly in terms of compatibility with traditional structural materials and systems. The aerospace industry is heavily reliant on established materials such as alloys of aluminum, magnesium and titanium, and transitioning to composite-based structures [246]. Integration of smart functionalities while maintaining mechanical properties is challenging [244, 245]. Carbon-based nanomaterials like graphene and CNTs show potential for enhancing composite properties, but their effective incorporation using scalable manufacturing techniques is difficult [244]. Though composites offer superior properties, they still face limitations such as insufficient mechanical properties and corrosion issues [239]. Ongoing research focuses on developing aerospace materials with enhanced mechanical performance and corrosion resistance.

6.6.4 Future Research Directions

Smart functional composite materials for aerospace holds exciting possibilities to address current limitations and advance innovation. Focal areas include exploring advanced nanomaterials like carbon nanotubes and graphene to enhance mechanical, electrical, and thermal properties [244]. Researchers are also developing multi-functional and autonomous systems, integrating self-healing, self-sensing, and energy-harvesting capabilities to create materials that adapt to environmental changes independently [247]. Future research directions also include the advancement of temperature control, fire-resistant materials, and self-healing polymers solutions, and multifunctional carbon fiber composites to enhance the safety, durability, and efficiency of aerospace structures [240]. Sustainability is a growing priority, driving efforts to develop recyclable and biodegradable composites with extended service lives [248]. Additionally, smart coatings and interfaces that protect against damage, corrosion, and wear are gaining traction [249]. Lastly, artificial intelligence and machine learning are revolutionizing composite design and manufacturing, enabling better material behavior prediction, process optimization, and integration of smart features, paving the way for a new era of aerospace innovation [250].

6.7 Conclusion

The development and application of smart functional composite polymers represent a metamorphic achievement in aerospace and mechanical engineering. These materials offer lightweight, high-strength, and multifunctional solutions capable of meeting the rigorous demands of modern industries. Critical mechanical characteristics like tensile strength, stiffness, and toughness, contribute significantly to the efficiency, durability, and performance of composite materials under applied loads and external forces. Thermal properties, including thermal conductivity and stability, play a significant role in thermomechanical performance, influencing interfacial interactions and ensuring reliability in high-temperature applications. Additionally, electrical properties like conductivity and resistivity, along with multifunctional capabilities such as self-sensing and self-healing, enhance the versatility and functionality of these materials. In the aerospace sector, smart materials are transforming both structural components, such as fuselage and wings, and functional elements, including actuators and energy storage systems. Advanced thermal management systems, including self-healing coatings, address overheating and ensure operational safety. Innovations in manufacturing techniques—from traditional methods to cutting-edge additive manufacturing and nanotechnology-enabled processes—are driving greater design flexibility and precision. However, challenges such as optimizing interfacial properties, reducing costs, improving scalability, and achieving ideal homogenization remain. Addressing these limitations requires focused research on enhancing mechanical, thermal, and multifunctional properties, integrating energy-harvesting capabilities, and leveraging artificial intelligence for material and process optimization. As industries evolve, smart composite materials are poised to revolutionize aerospace and mechanical engineering, contributing to safer, more efficient, and sustainable solutions. By overcoming current challenges, these innovations can unlock unprecedented opportunities, reshaping the future of aerospace technology and engineering.

References

1. Dehart, D. and Griffin, S., Astronautics Laboratory Smart Structures/Skins Overviews. Presented at the *First Joint U.S./Japan Adaptive Conference on Structures*, Hawaii, USA, November 13–15, 1990, Edwards AFB, CA, USA: The 93523.
2. Liu, T., Wang, H., Zhang, J., Biomimetic fabrication of new bioceramics-introduced fibrous scaffolds: From physicochemical characteristics to *in vitro* biological properties. *Mater. Sci. Eng., C*, 94, 547–557, 2019, doi: 10.1016/j.msec.2018.09.063.
3. Rizvi, S.T.R., Dengel, A., Ahmed, S., A hybrid approach and unified framework for bibliographic reference extraction. *IEEE Access*, 8, 217231–217245, 2020, doi: 10.1109/ACCESS.2020.3044030.
4. Liu, J., Zhang, J., Shintake, J., Wang, Y., Liu, L., Gao, X., Stimuli-responsive smart materials and structures: Advances in synthesis, characterization, and applications. *Materials*, Special Issue MDPI, 14, 1, 121, 2023, [Accessed Dec 2024].
5. Okoronkwo, M.U., Design, processing, and characterization of functional, smart, and sustainable materials: Recent trends. *Sustainability*, 2023, Topical Collection: Advances in Smart Material Design, 22, 2, 221, MDPI, [Accessed Dec 2024].
6. Liu, C. and Shenoi, R.A., Aeroelastic analysis and optimization of smart airfoil structures. *J. Aerosp. Eng.*, 28, 6, 1–10, 2015.
7. Febvre, M., Rodriguez, J., Chesne, S., Collet, M., Artificial Intelligence for Active Vibration Control Optimization on Smart Structures. *Proceedings of the ASME Conference on Smart Materials, Adaptive Structures, and Intelligent Systems*, Austin, Texas, USA, September 11–13, 2023, DOI: 10.1115/SMASIS2023-110216.
8. Wang, W., Xiang, Y., Yu, J., Yang, L., Development and Prospect of Smart Materials and Structures for Aerospace Sensing Systems and Applications. *Sensors*, 23, 3, 1545, 2023, DOI: 10.3390/s23031545.
9. Kontiza, A. and Kartsonakis, I.A., Smart Composite Materials with Self-Healing Properties: A Review on Design and Applications. *Polymers*, 16, 2115, 2024, https://doi.org/10.3390/ polym16152115.
10. Yu, Y., Shi, Y., Kurita, H., Jia, Y., Wang, Z., Narita, F., Carbon Fiber-Reinforced Piezoelectric Nanocomposites: Design, Fabrication and Evaluation for Damage

Detection and Energy Harvesting. *Compos. Part A. Appl. Sci. Manuf.*, 172, 107587, 2023.
11. MohdJani, J., Leary, M., Subic, A., Gibson, M.A., A review of shape memory alloy research, applications and opportunities. *Mater. Des.*, 56, 1078–1113, 2014.
12. Kuilla, T., Bhadra, S., Yao, D., Kim, N.H., Bose, S., Lee, J.H., Recent advances in graphene-based polymer composites. *Prog. Polym. Sci.*, 35, 1350–1375, 2010.
13. Liu, X. and Li, Q., Advances in Smart Structures for Aerospace Applications: A Review. *Aerosp. Sci. Technol.*, 129, 108874, 2023, DOI: 10.1016/j.ast.2023.108874.
14. Rashid, T. and Ahmed, H., Smart Material Applications in Vibration Control for Aerospace Engineering. *Mater. Today: Proc.*, 72, 212–220, 2024, DOI: 10.1016/j.matpr.2024.01.003.
15. Manickam, T.S., Arockiasamy, F.S., Ponnusamy, B., Sivakumar, R., Sivakumar, P., Kandasamy, P., Comprehensive review of polymer composite energy storage applications. *Eng. Proc.*, *61*, 1, 24, 2024, DOI: 10.3390/engproc2024061024.
16. Perrin, H., Mertz, G., Senoussaoui, N.-L., Borghini, L., Klein, S., Vaudemont, R., Surface functionalization of thermoset composite for infrared hybrid welding. *Funct. Compos. Mater.*, 2, 10, 1–13, 2021, DOI: 10.1186/s42252-021-00010.
17. Suresh, G., *et al.*, Smart Polymer Nanocomposites: Recent Advances and Applications in Aerospace. *Prog. Aerosp. Sci.*, 135, 100809, 2023.
18. Dehart, D. and Griffin, S., Astronautics Laboratory Smart Structures/Skin Overviews. *First Joint U.S./Japan Adaptive Conference on Structures*, Hawaii, USA, Nov. 13–15, 1990, Edwards AFB, CA 93523, USA.
19. Smith, J. and Johnson, R., Innovations in Composite Materials for Aerospace Applications. *J. Aerosp. Eng.*, 25, 3, 145–156, 2021.
20. Gibson, R.F., *Principles of Composite Material Mechanics*, pp. 608, CRC Press, 2021. https://doi.org/10.1201/9781420014242
21. Strong, A.B., *Fundamentals of Composites Manufacturing: Materials, Methods, and Applications*, Society of Manufacturing Engineers, Dearborn, Mich, 2020.
22. Altin-Karataş, M. and Gökkaya, H., A review on machinability of carbon fiber reinforced polymer (CFRP) and glass fiber reinforced polymer (GFRP) composite materials. *Def. Technol.*, 14, 4, 318–326, 2018, doi:10.1016/j.dt.2018.02.001.
23. Kaw, A.K., *Mechanics of Composite Materials*, 2nd Edition, Boca Raton, CRC Press, 2022. https://doi.org/10.1201/9781420058291
24. Liu, X. and Li, Q., Advances in Composite Materials for Aerospace Applications: A Review. *Aerosp. Sci. Technol.*, 129, 108874, 2023, DOI: 10.1016/j.ast.2023.108874.
25. Smith, J. and Johnson, R., Corrosion Resistance of Composite Materials in Aerospace Applications. *J. Aerosp. Eng.*, 30, 2, 123–134, 2022.

26. Taylor, L. and Patel, M., Advancements in Composite Materials for Aircraft Structures. *Aerosp. Mater. Rev.*, 15, 4, 45–58, 2023.
27. Soutis, C. and Beaumont, P.W., *The Structural Integrity of Carbon Fiber Composites*, Springer, Springer International Publishing Switzerland, 2021.
28. Ahmad, Z. and Gibson, A.G., Composite Materials in Aerospace Engineering: Current Applications and Future Prospects. *J. Aerosp. Mater. Technol.*, *56*, 2, 145–165, 2024.
29. Brown, T. and Green, A., Design Flexibility of Composite Materials in Aerospace Applications. *Aerosp. Eng. J.*, 28, 3, 201–210, 2022.
30. Wilson, R. and Carter, S., Advancements in Composite Manufacturing for Aerodynamic Optimization. *J. Compos. Mater.*, 35, 5, 345–360, 2023.
31. Wiseman, D., Advantages of Composite Materials in the Aerospace Industry, 2018, https://medium.com/@danielcontend/advantages-of-composite-materials-in-the-aerospace-industry-f27c81f78c90.
32. Wang, Y. and Zhou, Y., Thermal and Electrical Applications of Composite Materials in Aerospace. *Mater. Today: Proc.*, *47*, 221–229, 2023, DOI: 10.1016/j.matpr.2023.06.001.
33. Scarponi, C., Carbon Fiber Composites in Aerospace Engineering, in: *Advanced Composite Materials for Aerospace Engineering, Processing, Properties and Applications*. pp. 385–412, Woodhead Publishing, Properties and Applications, Woodhead Publishing, 2022, 385-412, 2022, https://doi.org/10.1016/B978-0-08-100037-3.00013-4.
34. Soutis, C., The Role of Smart Composite Materials in Aerospace Structural Health Monitoring. *Smart Mater. Struct.*, *33*, 8, 085001, 2024, DOI: 10.1088/1361-665X/acdef9.
35. Martinez, L. and Thompson, J., Smart Materials in Aerospace: Enhancing Structural Health Monitoring. *J. Aerosp. Eng.*, 32, 4, 299–310, 2023.
36. Roberts, K. and Lee, M., Integration of Piezoelectric Fibers in Composite Materials for Real-Time Monitoring. *Compos. Sci. Technol.*, 40, 2, 150–162, 2022.
37. Johnson, R. and Smith, A., Advancements in Composite Manufacturing Techniques for Aerospace Applications. *J. Compos. Mater.*, 38, 6, 455–467, 2023.
38. Taylor, M. and Nguyen, H., Cost-Effectiveness of Automated Fiber Placement and Additive Manufacturing in Composite Production. *Aerosp. Manuf. Rev.*, 27, 3, 112–124, 2022.
39. Green, T. and Foster, L., The Role of Composite Materials in Reducing Fuel Consumption and CO_2 Emissions in Aerospace. *J. Sustain. Aviat.*, 15, 2, 87–98, 2023.
40. Carter, J. and Lee, S., Developments in Recyclable Composites for Aerospace Applications. *Compos. Sustain.*, 22, 1, 34–45, 2022.
41. Sahayaraj, A.F., Dhamotharan, S., Sandeep, D., Ramachandran, P., Jenish, I., Divakaran, D., Suyambulingam, I., Sanjay, M.R., Siengchin, S., Sustainable

Smart Polymer Composite Materials: A Comprehensive Review. *E3S Web Conf.*, 65, 02014, 2023.
42. Soutis, C., Smart Composite Materials: Advances and Applications in Aerospace Engineering. *Smart Mater. Struct.*, *31*, 3, 045001, 2022, DOI: 10.1088/1361-665X/acbc55.
43. Liu, Y. and Zhang, Y., Functional Polymer Composites: Innovations in Smart Applications. *Compos. Sci. Technol.*, *210*, 109758, 2023, DOI: 10.1016/j.compscitech.2023.109758.
44. Liu, Y. and Zhang, Y., Smart Composites in Biomedical Applications: Drug Delivery and Tissue Engineering. *J. Biomed. Mater. Res.*, *72*, 4, 1111–1124, 2023, DOI: 10.1002/jbm.24982.
45. Zhang, Y. and Wang, X., Smart Polymer Composites in Aerospace Engineering: Challenges and Innovations. *Aerosp. Sci. Technol.*, *105*, 105975, 2022, DOI: 10.1016/j.ast.2020.105975.
46. Cao, L., Liu, Y., Wang, Y., Tan, M., Zhang, H., Zhao, H., Zhang, Z., Xie, T., Wang, X., Zhou, L., Recent progress in smart polymer composite materials for responsive applications: A review. *Compos. Sci. Technol.*, 203, 108609, 2021.
47. Hu, J., Meng, H., Li, G., Ibekwe, S.I., A review of stimuli-responsive polymers for smart textile applications. *Smart Mater. Struct.*, 21, 5, 053001, 2012.
48. Ramakrishna, S., Mayer, J., Wintermantel, E., Leong, K.W., Biomedical applications of polymer-composite materials: A review. *Compos. Sci. Technol.*, 61, 9, 1189–1224, 2001.
49. Zhang, Z., Liu, Y., Leng, J., Recent advances in shape memory polymers and composites: A review. *J. Mater. Chem. A*, 2, 15, 4873–4885, 2014.
50. Awode, E.I., Amankwah, S., Mbada, N.I., Omiogbemi, I.M.B., Simulating lightning effects on carbon fiber composite shielded with carbon nanotube sheets using numerical methods. *Heliyon*, 10, 8, e29762, 2024, https://doi.org/10.1016/j.heliyon.2024.e29762.
51. Dhanasekar, S., Stella, T., Jothy, T., Thenmozhi, A., Bharathi, N., Divya, T., Thiyagarajan, K., Singh, P.R., Yanala, S.S., Ganganagunta, J., Jayakumar, M., Study of Polymer Matrix Composites for Electronics Applications. *J. Mater. Sci.*, 57, 10, 1–15, 2022.
52. Sajan, S. and Selvaraj, D.P., A Review on Polymer Matrix Composite Materials and Their Applications. *Materials*, 14, 2, 3913, 2021.
53. Hanmanthu, N.R. and Nikhil, G., Overview of Manufacturing PMC's. *Int. J. Recent Res. Civ. Mech. Eng. (IJRRCME)*, 6, 2, 4–7, 2020.
54. Chawla, K.K., *Composite Materials*, Springer Nature Switzerland AG, 1-560, 2019, doi:10.1007/978-3-030-28983-6.
55. Soutis, C., Carbon fiber reinforced plastics in aircraft construction. *Mater. Sci. Eng. A*, 412, 1–2, 171–176, 2005, doi:10.1016/j.msea.2005.08.064.
56. Khan, S.U., Kim, J.K., Marom, G., Carbon fiber reinforced composites: An overview of recent innovations and future trends. *Compos. Sci. Technol.*, 194, 108174, 2020.

57. Deng, S., Djukic, L., Paton, R., Ye, L., Thermoplastic–thermoset hybrid composites: A review. *Composites, Part A: Appl. Sci. Manuf.*, 119, 100–117, 2019.
58. Li, Z., Zhang, Z., Li, K., Zhang, Y., Liu, Y., Leng, J., High-performance carbon fiber reinforced polymer composites for aerospace applications: A review. *Compos. Commun.*, 31, 101095, 2022.
59. Saeedifar, M. and Zarouchas, D., Damage characterization of carbon fiber-reinforced polymer composites under cyclic loading: A review. *Compos. Part B: Eng.*, 195, 108039, 2020.
60. M'Saoubi, R., Axinte, D., Soo, S.L., Nobel, C., Attia, H., Kappmeyer, G., Engin, S., Sim, W.M., High performance cutting of advanced aerospace alloys and composite materials. *CIRP Ann. – Manuf. Technol.*, 64, 2, 557–580, 2015.
61. Rajak, D.K., Wagh, P.H., Linul, E., Manufacturing Technologies of Carbon/Glass Fiber-Reinforced Polymer Composites and Their Properties: A Review. *Polymers, 13*, 21, 3721, 2021, https://doi.org/10.3390/polym13213721.
62. Erhard, G. and Thompson, M., *Designing with Plastics*, Hanser, Munich, Germany, 2006.
63. Mangalgiri, P.D., Composite materials for aerospace applications. *Bull. Mater. Sci.*, 22, 657–664, 1999.
64. Ho, K.C. and Jeng, M.C., Tribological characteristics of short glass fibre reinforced polycarbonate composites. *Wear*, 206, 60–68, 1997.
65. Jendli, Z., Fitoussi, J., Meraghni, F., Baptiste, D., Anisotropic strain rate effects on the fibre-matrix interface decohesion in sheet moulding compound composites. *Compos. Sci. Technol.*, 65, 387–393, 2005.
66. Anand, P., Rajesh, D., Senthil Kumar, M., Raj, I.S., Investigations on the performances of treated jute/Kenaf hybrid natural fiber reinforced epoxy composite. *J. Polym. Res.*, 25, 94, 2018.
67. Sathishkumar, T., Satheeshkumar, S., Naveen, J., Glass fiber-reinforced polymer composites – a review. *J. Reinf. Plast. Compos.*, 33, 13, 1258–1275, 2014, doi:10.1177/0731684414530790.
68. Drummer, J., Felwa, T., Bodo, F., Effect of Fiber Misalignment and Environmental Temperature on the Compressive Behavior of Fiber Composites. *Polymers*, 15, 13, 2833, 2023.
69. Khatua, V., Gurumoorthy, B., Ananthasuresh, G.K., Selectively Embedding Multiple Spatially Steered Fibers in Polymer Composite Parts Made Using Vat Photopolymerization. *Software Condens. Matte.*, 23, 3, 15813, 2023, arXiv preprint arXiv.
70. Zhang, Y. and Zhang, X., Advancements in Fiber-Reinforced Polymer (FRP) Composites. *Front. Mater.*, 11, 91, 2024.
71. Dayyoub, T., Maksimkin, A.V., Filippova, O.V., Tcherdyntsev, V.V., Telyshev, D.V., Shape Memory Polymers as Smart Materials: A Review. *Polymers*, 14, 17, 3511, 2022, https://doi.org/10.3390/polym14173511.
72. Menon, A.V., Madras, G., Bose, S., The Journey of Self-Healing and Shape Memory Polyurethanes from Bench to Translational Research. *Polym. Chem.*, 10, 4370–4388, 2019.

73. Jamari, J., Ammarullah, M.I., Santoso, G., Sugiharto, S., Supriyono, T., Prakoso, A.T., Basri, H., van der Heide, E., Computational Contact Pressure Prediction of CoCrMo, SS 316L and Ti6Al4V Femoral Head against UHMWPE Acetabular Cup under Gait Cycle. *J. Funct. Biomater.*, 13, 64, 2022.
74. Lendlein, A., Jiang, H.Y., Junger, O., Langer, R., Light-induced shape-memory polymers. *Nature*, 434, 879–882, 2005.
75. Agarwal, S., Jiang, S., Chen, Y., Progress in the Field of Water- and/or Temperature-Triggered Polymer Actuators. *Macromol. Mater. Eng.*, 304, 1800548, 2018.
76. Chen, Y., Chen, C., Rehman, H.U., Zheng, X., Li, H., Liu, H., Hedenqvist, M.S., Shape-Memory Polymeric Artificial Muscles: Mechanisms, Applications and Challenges. *Molecules*, 25, 4246, 2020.
77. Zare, M., Prabhakaran, M.P., Parvin, N., Ramakrishna, S., Thermally-induced Two-way Shape Memory Polymers: Mechanisms, Structures, and Applications. *Chem. Eng. J.*, 374, 706–720, 2019.
78. Sarikaya, S., Gardea, F., Auletta, J.T., Kavosi, J., Langrock, A., Mackie, D.M., Naraghi, M., Athermal artificial muscles with drastically improved work capacity from pH-Responsive coiled polymer fibers. *Sens. Actuators B Chem.*, 335, 129703, 2021.
79. Chaudhary, V., Punia Bangar, S., Thakur, N., Trif, M., Recent Advancements in Smart Biogenic Packaging: Reshaping the Future of the Food Packaging Industry. *Polymers*, 14, 829, 2022.
80. Yazik, M.H.M. and Sultan, M.T.H., Shape memory polymer and its composites as morphing materials, in: *Failure Analysis in Biocomposites, Fibre-Reinforced Composites and Hybrid Composites*, 1st ed., M. Jawaid, M. Mohamed Thariq, N. Saba (Eds.), pp. 181–198, Woodhead Publishing, Amsterdam, The Netherlands, 2019.
81. Hornat, C.C. and Urban, M.W., Shape Memory Effects in Self-Healing Polymers. *Prog. Polym. Sci.*, 102, 101208, 2020, doi:10.1016/j.progpolymsci.2020.101208.
82. Urban, M.W., *Stimuli-responsive materials: from molecules to nature mimicking materials design*, p. 475, Royal Society of Chemistry, Cambridge, 2016.
83. Sumerlin, B.S., Next-generation self-healing materials. *Science*, 362, 150–161, 2018.
84. White, S.R., Sottos, N.R., Geubelle, P.H., Moore, J.S., Autonomic healing of polymer composites. *Nature*, 409, 794–7, 2001.
85. Park, J.H. and Braun, P.V., Coaxial electrospinning of self-healing coatings. *Adv. Mater.*, 22, 496–9, 2010.
86. Sinha-Ray, S., Pelot, D.D., Zhou, Z.P., Rahman, A., Wu, X.F., Yarin, A.L., Encapsulation of self-healing materials by coelectrospinning, emulsion electrospinning, solution blowing and intercalation. *J. Mater. Chem.*, 22, 9138–9146, 2014.

87. Lee, M.W., An, S., Lee, C., Liou, M., Yarin, A.L., Yoon, S.S., Self-healing transparent core–shell nanofiber coatings for anti-corrosive protection. *J. Mater. Chem. A*, 2, 7045–7053, 2014.
88. Li, C.H., Wang, C., Keplinger, C., Zuo, J.L., Jin, L., Sun, Y., *et al.*, A highly stretchable autonomous self-healing elastomer. *Nat. Chem.*, 8, 6, 618–624, 2016.
89. Yang, Y. and Urban, M.W., Self-healing of glucose-modified polyurethane networks facilitated by damage-induced primary amines. *Polym. Chem.*, 8, 303–9, 2017.
90. An, N., Wang, X., Li, Y., Zhang, L., Lu, Z., Sun, J., Healable and mechanically super-strong polymeric composites derived from hydrogen-bonded polymeric complexes. *Adv. Mater.*, 31, 1904882/1–7, 2019.
91. Huang, L., Yi, N., Wu, Y., Zhang, Y., Zhang, Q., Huang, Y., *et al.*, Multichannel and repeatable self-healing of mechanical enhanced graphene-thermoplastic polyurethane composites. *Adv. Mater.*, 25, 2224–8, 2013.
92. Urban, M.W., Davydovich, D., Yang, Y., Demir, T., Zhang, Y., Casabianca, L., Key-and-lock commodity self-healing copolymers. *Science*, 362, 220–5, 2018.
93. Yang, Y., Ding, X., Urban, M.W., Chemical and physical aspects of self-healing materials. *Prog. Polym. Sci.*, 49, 34–59, 2015.
94. Xie, T., Tunable polymer multi-shape memory effect. *Nature*, 464, 267–70, 2010.
95. Thongruang, W., Spontak, R.J., Balik, C.M., Bridged double percolation in conductive polymer composites: an electrical conductivity, morphology and mechanical property study. *Polymer*, 43, 3717e25, 2002.
96. Feller, J.F., Chauvelon, P., Linossier, I., Glouannec, P., Characterization of electrical and thermal properties of extruded tapes of thermoplastic conductive polymer composites (CPC). *Polym. Test.*, 22, 831e7, 2003.
97. Petrach, E., Abu-Isa, I., Wang, X., Percolation threshold study of a plasticeelastomeric matrix based composite material for bipolar plates in proton exchange membrane fuel cells. *J. Compos. Mater.*, 46, 2959e71, 2012.
98. Mohd, R.N.A., Sulong, A.B., Sahari, J., A review of electrical conductivity models for conductive polymer composite. *Int. J. Hydrogen Energy*, 42, 14, 9262–9273, 2017, doi:10.1016/j.ijhydene.2016.03.045.
99. Roh, H., Kim, D.-H., Cho, Y., Jo, Y.-M., del-Alamo, J.A., Kulik, H.J., Dincă, M., Gumyusenge, A., Robust chemiresistive behavior in conductive polymer/MOF composites. *Adv. Mater.*, *36*, 27, 2312382, 2024, https://doi.org/10.1002/adma.202312382.
100. Barbieri, E., Bilotti, E., Liu, Y., Grimaldi, C., Pyroresistive response of percolating conductive polymer composites. *Phys. Rev. Mater.*, 8, 4, 045602, 2024.
101. Henry, A., Thermal transport in polymers. *Annu. Rev. Heat Transfer*, 17, 485–520, 2013.
102. Tjong, S.C., Structural and mechanical properties of polymer nanocomposites. *Mater. Sci. Eng., R*, 53, 3, 73–197, 2006.

103. Bartczak, Z., Argon, A., Cohen, R., Weinberg, M., Toughness mechanism in semi-crystalline polymer blends: II. High-density polyethylene toughened with calcium carbonate filler particles. *Polymer*, 40, 9, 2347–2365, 1999.
104. Misra, R., Nerikar, P., Bertrand, K., Murphy, D., Some aspects of surface deformation and fracture of 5–20% calcium carbonatereinforced polyethylene composites. *Mater. Sci. Eng., A*, 384, 1, 284–298, 2004.
105. Zhu, J., Wei, S., Alexander, M.J., Cocke, D., Ho, T.C., Guo, Z., Electrical conductivity manipulation and switching phenomena of poly (p-phenylenebenzobisthiazole) thin film by doping process. *J. Mater. Chem.*, 20, 568–574, 2010.
106. Fu, S.-Y., Feng, X.-Q., Lauke, B., Mai, Y.-W., Effects of particle size, particle/matrix interface adhesion and particle loading on mechanical properties of particulate–polymer composites. *Compos. Part B*, 39, 6, 933–961, 2008.
107. Calvert, P., Nanotube composites: a recipe for strength. *Nature*, 399, 6733, 210–211, 1999.
108. Gojny, F., Wichmann, M., Köpke, U., Fiedler, B., Schulte, K., Carbon nanotube-reinforced epoxy-composites: enhanced stiffness and fracture toughness at low nanotube content. *Compos. Sci. Technol.*, 64, 15, 2363–2371, 2004.
109. Wang, X., Kalali, E.N., Wang, D.-Y., An *in situ* polymerization approach for functionalized MoS2/nylon-6 nanocomposites with enhanced mechanical properties and thermal stability. *J. Mater. Chem. A*, 3, 47, 24112–24120, 2015.
110. Sahoo, N.G., Rana, S., Cho, J.W., Li, L., Chan, S.H., Polymer nanocomposites based on functionalized carbon nanotubes. *Prog. Polym. Sci.*, 35, 7, 837–867, 2010.
111. Ou, Y., Yang, F., Yu, Z.Z., A new conception on the toughness of nylon 6/silica nanocomposite prepared *via in situ* polymerization. *J. Polym. Sci., Part B: Polym. Phys.*, 36, 5, 789–795, 1998.
112. Gong, T., Liu, M.-Q., Liu, H., Peng, S.-P., Li, T., Bao, R.-Y., Yang, W., Xie, B.-H., Yang, M.-B., Guo, Z., Selective distribution and migration of carbon nanotubes enhanced electrical and mechanical performances in polyolefin elastomers. *Polymer*, 110, 1–11, 2017.
113. Thostenson, E.T. and Chou, T.-W., Aligned multi-walled carbon nanotube-reinforced composites: processing and mechanical characterization. *J. Phys. D Appl. Phys.*, 35, 16, L77, 2002.
114. Shofner, M.L., Khabashesku, V.N., Barrera, E.V., Processing and mechanical properties of fluorinated single-wall carbon nanotube–polyethylene composites. *Chem. Mater.*, 18, 4, 906–913, 2006.
115. Qian, D., Dickey, E.C., Andrews, R., Rantell, T., Load transfer and deformation mechanisms in carbon nanotube-polystyrene composites. *Appl. Phys. Lett.*, 76, 20, 2868–2870, 2000.
116. Blake, R., Coleman, J.N., Byrne, M.T., McCarthy, J.E., Perova, T.S., Blau, W.J., Fonseca, A., Nagy, J.B., Gun'ko, Y.K., Reinforcement of poly (vinyl chloride) and polystyrene using chlorinated polypropylene grafted carbon nanotubes. *J. Mater. Chem.*, 16, 43, 4206–4213, 2006.

117. Xiao, K., Zhang, L., Zarudi, I., Mechanical and rheological properties of carbon nanotube-reinforced polyethylene composites. *Compos. Sci. Technol.*, 67, 2, 177–182, 2007.
118. Yang, B.X., Pramoda, K.P., Xu, G.Q., Goh, S.H., Mechanical reinforcement of polyethylene using polyethylene-grafted multiwalled carbon nanotubes. *Adv. Funct. Mater.*, 17, 13, 2062–2069, 2007.
119. Lee, J.-Y., Zhang, Q., Emrick, T., Crosby, A.J., Nanoparticle alignment and repulsion during failure of glassy polymer nanocomposites. *Macromolecules*, 39, 21, 7392–7396, 2006.
120. Hsiao, C.-C., Lin, T.S., Cheng, L.Y., Ma, C.-C.M., Yang, A.C.-M., The nanomechanical properties of polystyrene thin films embedded with surface-grafted multiwalled carbon nanotubes. *Macromolecules*, 38, 11, 4811–4818, 2005.
121. Stafford, C.M., Guo, S., Harrison, C., Chiang, M.Y., Combinatorial and high-throughput measurements of the modulus of thin polymer films. *Rev. Sci. Instrum.*, 76, 6, 062207, 2005.
122. Stafford, C.M., Harrison, C., Beers, K.L., Karim, A., Amis, E.J., VanLandingham, M.R., Kim, H.-C., Volksen, W., Miller, R.D., Simonyi, E.E., A buckling-based metrology for measuring the elastic moduli of polymeric thin films. *Nat. Mater.*, 3, 8, 545–550, 2004.
123. Stafford, C.M., Vogt, B.D., Harrison, C., Julthongpiput, D., Huang, R., Elastic moduli of ultrathin amorphous polymer films. *Macromolecules*, 39, 15, 5095–5099, 2006.
124. Dekkers, M. and Heikens, D., The effect of interfacial adhesion on the tensile behavior of polystyrene–glass–bead composites. *J. Appl. Polym. Sci.*, 28, 12, 3809–3815, 1983.
125. Fu, S.-Y. and Lauke, B., Characterization of tensile behaviour of hybrid short glass fibre/calcite particle/ABS composites. *Compos. Part A*, 29, 5, 575–583, 1998.
126. Zhu, Z.K., Yang, Y., Yin, J., Qi, Z.N., Preparation and properties of organosoluble polyimide/silica hybrid materials by sol–gel process. *J. Appl. Polym. Sci.*, 73, 14, 2977–2984, 1999.
127. Zhu, J., Mu, L., Chen, L., Shi, Y., Wang, H., Feng, X., Lu, X., Interface-strengthened polyimide/carbon nanofibers nanocomposites with superior mechanical and tribological properties. *Macromol. Chem. Phys.*, 215, 14, 1407–1414, 2014.
128. Q-l, X. and X-g, T., Atomistic modeling of mechanical characteristics of CNT-polyethylene with interfacial covalent interaction. *J. Nanomater.*, 237520, 1–9, 2015.
129. Ramanathan, T., Liu, H., Brinson, L., Functionalized SWNT/polymer nanocomposites for dramatic property improvement. *J. Polym. Sci. Part B Polym. Phys.*, 43, 17, 2269–2279, 2005.
130. Zhang, J., Jia, Z., Jia, D., Zhang, D., Zhang, A., Chemical functionalization for improving dispersion and interfacial bondingof halloysite nanotubes in epoxy nanocomposites. *High Perform. Polym.*, 26, 7, 734–743, 2014.

131. Thio, Y., Argon, A., Cohen, R., Role of interfacial adhesion strength on toughening polypropylene with rigid particles. *Polymer*, 45, 10, 3139–3147, 2004.
132. Wu, C.L., Zhang, M.Q., Rong, M.Z., Friedrich, K., Silica nanoparticles filled polypropylene: effects of particle surface treatment, matrix ductility and particle species on mechanical performance of the composites. *Compos. Sci. Technol.*, 65, 3, 635–645, 2005.
133. Frankland, S., Caglar, A., Brenner, D., Griebel, M., Molecular simulation of the influence of chemical cross-links on the shear strength of carbon nanotube–polymer interfaces. *J. Phys. Chem. B*, 106, 12, 3046–3048, 2002.
134. Gao, J., Itkis, M.E., Yu, A., Bekyarova, E., Zhao, B., Haddon, R.C., Continuous spinning of a single-walled carbon nanotube–nylon composite fiber. *JACS*, 127, 11, 3847–3854, 2005.
135. Liff, S.M., Kumar, N., McKinley, G.H., High-performance elastomeric nanocomposites *via* solvent-exchange processing. *Nat. Mater.*, 6, 1, 76–83, 2007.
136. Wang, K., Chen, L., Wu, J., Toh, M.L., He, C., Yee, A.F., Epoxy nanocomposites with highly exfoliated clay: mechanical properties and fracture mechanisms. *Macromolecules*, 38, 3, 788–800, 2005.
137. Friedrich, K., Fakirov, S., Zhang, Z., *Polymer composites: from nano-to macro-scale*, Springer Science & Business Media, New York, 2005.
138. Michler, G.H. and Balta-Calleja, F.J., *Mechanical properties of polymers based on nanostructure and morphology*, vol. 71, CRC Press, Florida, 2016.
139. Moloney, A., Kausch, H., Kaiser, T., Beer, H., Parameters determining the strength and toughness of particulate filled epoxide resins. *J. Mater. Sci.*, 22, 2, 381–393, 1987.
140. Ash, B.J., Siegel, R.W., Schadler, L.S., Mechanical behavior of alumina/poly (methyl methacrylate) nanocomposites. *Macromolecules*, 37, 4, 1358–1369, 2004.
141. Johnsen, B., Kinloch, A., Mohammed, R., Taylor, A., Sprenger, S., Toughening mechanisms of nanoparticle-modified epoxy polymers. *Polymer*, 48, 2, 530–541, 2007.
142. Blumstein, A., Polymerization of adsorbed monolayers. II. Thermal degradation of the inserted polymer. *J. Polym. Sci. Part A Polym. Chem.*, 3, 7, 2665–2672, 1965.
143. Gilman, J.W., Flammability and thermal stability studies of polymer layered-silicate (clay) nanocomposites. *Appl. Clay Sci.*, 15, 1, 31–49, 1999.
144. Pielichowski, K. and Leszczynska, A., Polyoxymethylene-based nanocomposites with montmorillonite: an introductory study. *Polimery*, 51, 2, 143–149, 2006.
145. Cao, Y., Lai, Z., Feng, J., Wu, P., Graphene oxide sheets covalently functionalized with block copolymers *via* click chemistry as reinforcing fillers. *J. Mater. Chem.*, 21, 25, 9271–9278, 2011.

146. Park, S.-J., Seo, D.-I., Lee, J.-R., Surface modification of montmorillonite on surface acid–base characteristics of clay and thermal stability of epoxy/clay nanocomposites. *J. Colloid Interface Sci.*, 251, 1, 160–165, 2002.
147. Kashiwagi, T., Grulke, E., Hilding, J., Groth, K., Harris, R., Butler, K., Shields, J., Kharchenko, S., Douglas, J., Thermal and flammability properties of polypropylene/carbon nanotube nanocomposites. *Polymer*, 45, 12, 4227–4239, 2004.
148. Kashiwagi, T., Du, F., Winey, K.I., Groth, K.M., Shields, J.R., Bellayer, S.P., Kim, H., Douglas, J.F., Flammability properties of polymer nanocomposites with single-walled carbon nanotubes: effects of nanotube dispersion and concentration. *Polymer*, 46, 2, 471–481, 2005.
149. Kashiwagi, T., Du, F., Douglas, J.F., Winey, K.I., Harris, R.H., Shields, J.R., Nanoparticle networks reduce the flammability of polymer nanocomposites. *Nat. Mater.*, 4, 12, 928–933, 2005.
150. Yadav, S.K. and Cho, J.W., Functionalized graphene nanoplatelets for enhanced mechanical and thermal properties of polyurethane nanocomposites. *Appl. Surf. Sci.*, 266, 360–367, 2013.
151. Zhu, J., Uhl, F.M., Morgan, A.B., Wilkie, C.A., Studies on the mechanism by which the formation of nanocomposites enhances thermal stability. *Chem. Mater.*, 13, 12, 4649–4654, 2001.
152. Hu, L., Jiang, P., Zhang, P., Bian, G., Sheng, S., Huang, M., Bao, Y., Xia, J., Amine-graphene oxide/waterborne polyurethane nanocomposites: effects of different amine modifiers on physical properties. *J. Mater. Sci.*, 51, 18, 8296–8309, 2016.
153. Salavagione, H.J., Gomez, M.A., Martínez, G., Polymeric modification of graphene through esterification of graphite oxide and poly (vinyl alcohol). *Macromolecules*, 42, 17, 6331–6334, 2009.
154. Liu, N., Luo, F., Wu, H., Liu, Y., Zhang, C., Chen, J., One-step ionic-liquid-assisted electrochemical synthesis of ionic-liquidfunctionalized graphene sheets directly from graphite. *Adv. Funct. Mater.*, 18, 10, 1518–1525, 2008.
155. Kim, H., Abdala, A.A., Macosko, C.W., Graphene/polymer nanocomposites. *Macromolecules*, 43, 16, 6515–6530236, 2010. Pettersson, S. and Mahan, G., Theory of the thermal boundary resistance between dissimilar lattices. *Phys. Rev. B*, 42, 12, 7386, 1990.
156. He, H., Fu, R., Shen, Y., Han, Y., Song, X., Preparation and properties of Si3N4/PS composites used for electronic packaging. *Compos. Sci. Technol.*, 67, 11, 2493–2499, 2007.
157. Norris, P.M., Le, N.Q., Baker, C.H., Tuning phonon transport: from interfaces to nanostructures. *J. Heat Transfer*, 135, 6, 061604, 2013.
158. Wang, Z., Carter, J.A., Lagutchev, A., Koh, Y.K., Seong, N.-H., Cahill, D.G., Dlott, D.D., Ultrafast flash thermal conductance of molecular chains. *Science*, 317, 5839, 787–790, 2007.
159. Nan, C.-W., Liu, G., Lin, Y., Li, M., Interface effect on thermal conductivity of carbon nanotube composites. *Appl. Phys. Lett.*, 85, 16, 3549–3551, 2004.

160. Kochetov, R., Korobko, A., Andritsch, T., Morshuis, P., Picken, S., Smit, J., Modelling of the thermal conductivity in polymer nanocomposites and the impact of the interface between filler and matrix. *J. Phys. D Appl. Phys.*, 44, 39, 395401, 2011, Adv Compos Hybrid Mater.
161. Costescu, R., Cahill, D., Fabreguette, F., Sechrist, Z., George, S., Ultra-low thermal conductivity in W/Al2O3 nanolaminates. *Science*, 303, 5660, 989–990, 2004.
162. Park, J.G., Cheng, Q., Lu, J., Bao, J., Li, S., Tian, Y., Liang, Z., Zhang, C., Wang, B., Thermal conductivity of MWCNT/epoxy composites: the effects of length, alignment and functionalization. *Carbon*, 50, 6, 2083–2090, 2012.
163. Li, J., Ma, P.C., Chow, W.S., To, C.K., Tang, B.Z., Kim, J.K., Correlations between percolation threshold, dispersion state, and aspect ratio of carbon nanotubes. *Adv. Funct. Mater.*, 17, 16, 3207–3215, 2007.
164. Deng, F., Zheng, Q.-S., Wang, L.-F., Nan, C.-W., Effects of anisotropy, aspect ratio, and nonstraightness of carbon nanotubes on thermal conductivity of carbon nanotube composites. *Appl. Phys. Lett.*, 90, 2, 021914, 2007.
165. Veca, L.M., Meziani, M.J., Wang, W., Wang, X., Lu, F., Zhang, P., Lin, Y., Fee, R., Connell, J.W., Sun, Y.P., Carbon nanosheets for polymeric nanocomposites with high thermal conductivity. *Adv. Mater.*, 21, 20, 2088–2092, 2009.
166. Burger, N., Laachachi, A., Ferriol, M., Lutz, M., Toniazzo, V., Ruch, D., Review of thermal conductivity in composites: mechanisms, parameters and theory. *Prog. Polym. Sci.*, 61, 1–28, 2016.
167. Smith, D.K. and Pantoya, M.L., Effect of nanofiller shape on effective thermal conductivity of fluoropolymer composites. *Compos. Sci. Technol.*, 118, 251–256, 2015.
168. Chen, L., Sun, Y.-Y., Xu, H.-F., He, S.-J., Wei, G.-S., Du, X.-Z., Lin, J., Analytic modeling for the anisotropic thermal conductivity of polymer composites containing aligned hexagonal boron nitride. *Compos. Sci. Technol.*, 122, 42–49, 2016.
169. Prasher, R., Evans, W., Meakin, P., Fish, J., Phelan, P., Keblinski, P., Effect of aggregation on thermal conduction in colloidal nanofluids. *Appl. Phys. Lett.*, 89, 14, 143119, 2006.
170. Kim, G., Ryu, S.H., Lee, J.-T., Seong, K.-H., Lee, J.E., Yoon, P.-J., Kim, B.-S., Hussain, M., Choa, Y.-H., Enhancement of thermal conductive pathway of boron nitride coated polymethylsilsesquioxane composite. *J. Nanosci. Nanotechnol.*, 13, 11, 7695–7700, 2013.
171. Yu, W., Fu, J., Chen, L., Zong, P., Yin, J., Shang, D., Lu, Q., Chen, H., Shi, L., Enhanced thermal conductive property of epoxy composites by low mass fraction of organic–inorganic multilayer covalently grafted carbon nanotubes. *Compos. Sci. Technol.*, 125, 90–99, 2016.
172. Yan, H., Tang, Y., Su, J., Yang, X., Enhanced thermal–mechanical properties of polymer composites with hybrid boron nitride nanofillers. *Appl. Phys. A Mater. Sci. Process.*, 114, 2, 331–337, 2014.

173. Wong, C. and Bollampally, R.S., Comparative study of thermally conductive fillers for use in liquid encapsulants for electronic packaging. *IEEE Trans. Adv. Packag.*, 22, 1, 54–59, 1999.
174. Teng, C.-C., Ma, C.-C.M., L.C.-H., Yang, S.-Y., Lee, S.-H., Hsiao, M.-C., Yen, M.-Y., Chiou, K.-C., Lee, T.-M., Thermal conductivity and structure of non-covalent functionalized graphene/epoxy composites. *Carbon*, 49, 15, 5107–5116, 2011.
175. Sun, R., Yao, H., Zhang, H.-B., Li, Y., Mai, Y.-W., Yu, Z.-Z., Decoration of defect-free graphene nanoplatelets with alumina for thermally conductive and electrically insulating epoxy composites. *Compos. Sci. Technol.*, 137, 16–23, 2016.
176. Qian, R., Yu, J., Wu, C., Zhai, X., Jiang, P., Alumina-coated graphene sheet hybrids for electrically insulating polymer composites with high thermal conductivity. *RSC Adv.*, 3, 38, 17373–17379, 2013.
177. Gu, J., Xie, C., Li, H., Dang, J., Geng, W., Zhang, Q., Thermal percolation behavior of graphene nanoplatelets/polyphenylene sulfide thermal conductivity composites. *Polym. Compos.*, 35, 6, 1087–1092, 2014.
178. Chen, J., Chen, X., Meng, F., Li, D., Tian, X., Wang, Z., Zhou, Z., Super-high thermal conductivity of polyamide-6/graphenegraphene oxide composites through *in situ* polymerization. *High Perform. Polym.*, 29, 5, 585–594, 2017.
179. Dai, W., Yu, J., Liu, Z., Wang, Y., Song, Y., Lyu, J., Bai, H., Nishimura, K., Jiang, N., Enhanced thermal conductivity and retained electrical insulation for polyimide composites with SiC nanowires grown on graphene hybrid fillers. *Compos. Part A*, 76, 73–81, 2015.
180. Zong, P., Fu, J., Chen, L., Yin, J., Dong, X., Yuan, S., Shi, L., Deng, W., Effect of aminopropylisobutyl polyhedral oligomeric silsesquioxane functionalized graphene on the thermal conductivity and electrical insulation properties of epoxy composites. *RSC Adv.*, 6, 13, 10498–10506, 2016.
181. Cho, E.-C., Huang, J.-H., Li, C.-P., Chang-Jian, C.-W., Lee, K.-C., Hsiao, Y.-S., Huang, J.-H., Graphene-based thermoplastic composites and their application for LED thermal management. *Carbon*, 102, 66–73, 2016.
182. Ma, W.-S., Wu, L., Yang, F., Wang, S.-F., Non-covalently modified reduced graphene oxide/polyurethane nanocomposites with good mechanical and thermal properties. *J. Mater. Sci.*, 49, 2, 562–571, 2014.
183. Tanimoto, M., Yamagata, T., Miyata, K., Ando, S., Anisotropic thermal diffusivity of hexagonal boron nitride-filled polyimide films: effects of filler particle size, aggregation, orientation, and polymer chain rigidity. *ACS Appl. Mater. Interfaces*, 5, 10, 4374–4382, 2013.
184. Evans, W., Prasher, R., Fish, J., Meakin, P., Phelan, P., Keblinski, P., Effect of aggregation and interfacial thermal resistance on thermal conductivity of nanocomposites and colloidal nanofluids. *Int. J. Heat Mass Transfer*, 51, 5, 1431–1438, 2008.

185. Hsieh, W.-P., Lyons, A.S., Pop, E., Keblinski, P., Cahill, D.G., Pressure tuning of the thermal conductance of weak interfaces. *Phys. Rev. B*, 84, 18, 184107, 2011.
186. Gojny, F.H., Wichmann, M.H., Fiedler, B., Kinloch, I.A., Bauhofer, W., Windle, A.H., Schulte, K., Evaluation and identification of electrical and thermal conduction mechanisms in carbon nanotube/epoxy composites. *Polymer*, 47, 6, 2036–2045, 2006.
187. Zhu, J., Wei, S., Ryu, J., Budhathoki, M., Liang, G., Guo, Z., *In situ* stabilized carbon nanofiber (CNF) reinforced epoxy nanocomposites. *J. Mater. Chem.*, 20, 4937–4948, 2010.
188. Hu, N., Fukunaga, H., Atobe, S., Liu, Y., Li, J., Piezoresistive strain sensors made from carbon nanotubes based polymer nanocomposites. *Sensors*, 11, 11, 10691–10723, 2011.
189. Sandler, J., Kirk, J., Kinloch, I., Shaffer, M., Windle, A., Ultralow electrical percolation threshold in carbon-nanotube-epoxy composites. *Polymer*, 44, 19, 5893–5899, 2003.
190. Basheer, A.A., Advances in the smart materials applications in the aerospace industries. *Aircr. Eng. Aerosp. Technol.*, 92, 7, 1027–1035, 2020, https://doi.org/10.1108/AEAT-02-2020-0040.
191. Sudhakar, N., *Mechanics of Aero-structures*, Cambridge University Press & Assessment, Cambridge, CB2 8RU, UK, 2024, 978-1-107-07577-1.
192. NASA Glenn: Enabling future aerospace missions and new technological capabilities for NASA and the nation with our research, technology development, and engineering system development, 2024.
193. Edward, B., Abhinav, S., Prasun, B., Kai, F., Goebel, P., Stoelting, S.C., A Diagnostic Approach for Electro-Mechanical Actuators in Aerospace Systems, 2009, 978-1-4244-2622-5/09/$25.00 ©2009 IEEE. IEEEAC paper 1345.
194. Prosser, S.J., L.uca.vAdvanced Engmeertng Centre, Dog Kennel Lane, Shirley, Solrhull, West Mdiands, B90 4JJ UK, 1993.
195. Das, R., Melchior, C., Karumbaiah, K.M., Advanced Composite Materials for Aerospace Engineering. *Processing, Properties and Applications*, pp. 333–364, 2016. https://doi.org/10.1016/B978-0-08-100037-3.00011-0
196. Kausar, A., Ahmad, I., Rakha, S.A., Eisa, M.H., Diallo, A., State-Of-The-Art of Sandwich Composite Structures: Manufacturing—to—High Performance Applications. *J. Compos. Sci.* [Internet], 7, 3, 102, 2023, Available from: https://www.mdpi.com/2504-477X/7/3/102.
197. Chen, Y., Zhang, J., Li, Z., Zhang, H., Chen, J., Yang, W., *et al.*, Manufacturing Technology of Lightweight Fiber-Reinforced Composite Structures in Aerospace: Current Situation and toward Intellectualization. *Aerospace* [Internet], 10, 3, 206, 2023, Available from: https://www.mdpi.com/2226-4310/10/3/206.
198. Jiang, Y., Serrano, A.X., Choi, W., Advincula, R.C., Wu, H.F., Advanced and functional composite materials *via* additive manufacturing: Trends

and perspectives. *MRS Commun.*, 14, 4, 449–59, 2024 Aug 6, https://link.springer.com/10.1557/s43579-024-00625-5.
199. Pokharel, A., Falua, K.J., Babaei-Ghazvini, A., Acharya, B., Biobased Polymer Composites: A Review. *J. Compos. Sci.*, 6, 9, 255, 2022 Sep 5, https://www.mdpi.com/2504-477X/6/9/255.
200. Jayasekara, D., Lai, N.Y.G., Wong, K.H., Pawar, K., Zhu, Y., Level of automation (LOA) in aerospace composite manufacturing: Present status and future directions towards industry 4.0. *J. Manuf. Syst.*, 62, January, 44–61, 2022 Jan, https://linkinghub.elsevier.com/retrieve/pii/S0278612521002223.
201. Agwa, M., Youssef, S.M., Ali-Eldin, S.S., Megahed, M., Integrated vacuum assisted resin infusion and resin transfer molding technique for manufacturing of nano-filled glass fiber reinforced epoxy composite. *J. Ind. Text.*, 51, 3_suppl, 5113S–5144S, 2022 Jun 17, https://journals.sagepub.com/doi/10.1177/1528083720932337.
202. Sałasińska, K., Cabulis, P., Kirpluks, M., Kovalovs, A., Kozikowski, P., Barczewski, M., *et al.*, The Effect of Manufacture Process on Mechanical Properties and Burning Behavior of Epoxy-Based Hybrid Composites. *Materials*, 15, 1, 301, 2022 Jan 1, https://www.mdpi.com/1996-1944/15/1/301.
203. Herath, M. and Epaarachchi, J., Shape memory polymer composites and their smart structural applications, in: *Composite Materials*, pp. 581–610, Elsevier, Netherlands, 2021, https://linkinghub.elsevier.com/retrieve/pii/B9780128205129000010.
204. Middleton, B., Composites: Manufacture and Application., in: *Design and Manufacture of Plastic Components for Multifunctionality.* p. 53–101, Elsevier, 2016. https://doi.org/10.1016/C2014-0-00223-7
205. Yuhazri, M. and Sihombing, H.A., Comparison Process Between Vacuum Infusion and Hand Lay-Up Method Toward Kenaf/Polyester Composites. *Int. J. Basic Appl. Sci.*, 10, 3, 63–6, 2018.
206. Obertscheider, C. and Fauster, E., RTMsim - A Julia module for filling simulations in Resin Transfer Moulding with the Finite Area Method. *J. Open Source Softw.*, 8, 84, 4763, 2023 Apr 5, https://joss.theoj.org/papers/10.21105/joss.04763.
207. Miranda, Campos, B., Bourbigot, S., Fontaine, G., Bonnet, F., Thermoplastic matrix-based composites produced by resin transfer molding: A review. *Polym. Compos.*, 43, 5, 2485–506, 2022 May 11, https://4spepublications.onlinelibrary.wiley.com/doi/10.1002/pc.26575.
208. Kim, R.W., Kim, C.M., Hwang, K.H., Kim, S.R., Embedded Based Real-Time Monitoring in the High-Pressure Resin Transfer Molding Process for CFRP. *Appl. Sci.*, 9, 9, 1795, 2019 Apr 29, https://www.mdpi.com/2076-3417/9/9/1795.
209. Govignon, Q., Bickerton, S., Kelly, P. Liquid composite molding processes., in: *Advanced Fiber-Reinforced Polymer (FRP) Composites for Structural Applications.* pp. 101–36 Woodhead Publishing Series in Civil and Structural Engineering, Elsevier, 2023. https://doi.org/10.1016/B978-0-12-820346-0.00006-X.

210. Kerst, C.F. and Cutkosky, M.R., Compression Molding and Nickel Molds for Directional Gecko-Inspired Adhesives. *J. Micro Nano-Manuf.*, 9, 2, 7–10, 2021 Jun 1, https://asmedigitalcollection.asme.org/micronanomanufacturing/article/9/2/024501/1109232/Compression-Molding-and-Nickel-Molds-for.
211. Alnersson, G., Tahir, M.W., Ljung, A.L., Lundström, T.S., Review of the Numerical Modeling of Compression Molding of Sheet Molding Compound. *Processes*, 8, 2, 179, 2020 Feb 5, https://www.mdpi.com/2227-9717/8/2/179.
212. Lee, S., Shin, D., Kim, G., Ji, W., Numerical model for compression molding process of hybridly laminated thermoplastic composites based on anisotropic rheology. *Compos. Part C Open Access*, 7, March, 100215, 2022 Mar, https://linkinghub.elsevier.com/retrieve/pii/S2666682021001079.
213. Jiang, T., Zeng, G., Hu, C., Fabrication of highly filled wood plastic composite pallets with extrusion-compression molding technique. *Polym. Compos.*, 41, 7, 2724–31, 2020 Jul 14, https://4spepublications.onlinelibrary.wiley.com/doi/10.1002/pc.25570.
214. Rajeshkumar, G., Balaji, B.A., Seshadri, S.A., Compression and injection molding techniques, in: *Cellul Fibre Reinf Compos Interface Eng Process Perform.*, pp. 165–81, 2022.
215. Ali, N.I.M., Ali, M.A.M., Maidin, S., Sulaiman, M.A., Salleh, M.S., Bakar, M.H.A., Review on Experimental Design, Process Parameters and Responses of Compression Moulding Process, in: *Lecture Notes in Mechanical Engineering*, pp. 407–14, 2022, https://link.springer.com/10.1007/978-981-16-8954-3_38.
216. Ikram, H., Al Rashid, A., Koç, M., Additive manufacturing of smart polymeric composites: Literature review and future perspectives. *Polym. Compos.*, 43, 9, 6355–80, 2022 Sep 12, https://4spepublications.onlinelibrary.wiley.com/doi/10.1002/pc.26948.
217. Wu, H., Fahy, W.P., Kim, S., Kim, H., Zhao, N., Pilato, L., *et al.*, Recent developments in polymers/polymer nanocomposites for additive manufacturing. *Prog. Mater Sci.*, 111, June, 100638, 2020 Jun, https://linkinghub.elsevier.com/retrieve/pii/S0079642520300025.
218. López-Barroso, J., Flores-Hernández, C.G., Martínez-Hernández, A.L., Martínez-Barrera, G., Velasco-Santos, C. Additive Manufacturing for Complex Geometries in Polymer Composites. , in: *Advanced Composites*. pp. 121–86, Cham, Springer Nature Switzerland, 2023. https://doi.org/10.1007/978-3-031-42731-2_5
219. Xia, Y., He, Y., Zhang, F., Liu, Y., Leng, J.A., Review of Shape Memory Polymers and Composites: Mechanisms, Materials, and Applications. *Adv. Mater.*, 33, 6, 1–33, 2020 Feb 23, https://onlinelibrary.wiley.com/doi/10.1002/adma.202000713.
220. Zhu, W., Wang, N., Zhi, Q., Zhang, Z., Fu, L., Zhang, Y., *et al.*, Highly protective and functional strengthening strategies for 3D printed continuous carbon fiber reinforced polymer composites: Manufacturing and properties. *Compos. Commun.*, 52, December, 102145, 2024 Dec, https://linkinghub.elsevier.com/retrieve/pii/S245221392400336X.

221. Shepa, I., Mudra, E., Dusza, J., Electrospinning through the prism of time. *Mater. Today Chem.*, 21, 4, 100543, 2021 Aug, http://www.mediasphera.ru/issues/vestnik-otorinolaringologii/2021/4/1004246682021041095.
222. Zare, M., Dziemidowicz, K., Williams, G.R., Ramakrishna, S., Encapsulation of Pharmaceutical and Nutraceutical Active Ingredients Using Electrospinning Processes. *Nanomaterials*, 11, 8, 1968, 2021 Jul 30, https://www.mdpi.com/2079-4991/11/8/1968.
223. Tan, S.M., Teoh, X.Y., Le, Hwang, J., Khong, Z.P., Sejare, R., Almashhadani, A.Q., *et al.*, Electrospinning and its potential in fabricating pharmaceutical dosage form. *J. Drug Delivery Sci. Technol.*, 76, October, 103761, 2022 Oct, https://linkinghub.elsevier.com/retrieve/pii/S1773224722006724.
224. Al-Dhahebi, A.M., Ling, J., Krishnan, S.G., Yousefzadeh, M., Elumalai, N.K., Saheed, M.S.M., *et al.*, Electrospinning research and products: The road and the way forward. *Appl. Phys. Rev.*, 9, 1, 1–62, 2022 Mar 1, https://pubs.aip.org/apr/article/9/1/011319/2835574/Electrospinning-research-and-products-The-road-and.
225. Ahmadi, Bonakdar, M., Rodrigue, D., Electrospinning: Processes, Structures, and Materials. *Macromol.*, 4, 1, 58–103, 2024.
226. Böckl, B., Wedel, A., Misik, A., Drechsler, K., Effects of defects in automated fiber placement laminates and its correlation to automated optical inspection results. *J. Reinf. Plast. Compos.*, 42, 1–2, 3–16, 2023 Jan 5, https://journals.sagepub.com/doi/10.1177/07316844221093273.
227. Islam, F., Wanigasekara, C., Rajan, G., Swain, A., Prusty, B.G., An approach for process optimisation of the Automated Fibre Placement (AFP) based thermoplastic composites manufacturing using Machine Learning, photonic sensing and thermo-mechanics modelling. *Manuf. Lett.*, 32, April, 10–4, 2022 Apr, https://linkinghub.elsevier.com/retrieve/pii/S2213846322000025.
228. Air, A., Shamsuddoha, M., Gangadhara Prusty, B., A review of Type V composite pressure vessels and automated fibre placement based manufacturing. *Compos. Part B Eng.*, 253, March, 110573, 2023 Mar, https://linkinghub.elsevier.com/retrieve/pii/S1359836823005309.
229. Han, Z., Sun, S., Fu, H., Fu, Y., Multi-scale low-entropy method for optimizing the processing parameters during automated fiber placement. *Materials (Basel)*, 10, 9, 1–18, 2017.
230. Rashid, A.B., Haque, M., Islam, S.M.M., Uddin Labib, K.M.R., Nanotechnology-enhanced fiber-reinforced polymer composites: Recent advancements on processing techniques and applications. *Heliyon*, 10, 2, e24692, 2024 Jan, https://doi.org/10.1016/j.heliyon.2024.e24692.
231. Colorado, H.A., Gutierrez-Velasquez, E.I., Gil, L.D., *et al.*, Exploring the advantages and applications of nanocomposites produced *via* vat photopolymerization in additive manufacturing: A review. *Adv. Compos. Hybrid Mater.*, 7, 1, 2024, https://doi.org/10.1007/s42114-023-00808-z.
232. She, Y., Tang, J., Wang, C., Wang, Z., Huang, Z., Yang, Y., Nano-Additive Manufacturing and Non-Destructive Testing of Nanocomposites.

Nanomaterials, 13, 20, 2741, 2023 Oct 10, https://www.mdpi.com/2079-4991/13/20/2741.
233. Kaynan, O., Hosseini, E., Zakertabrizi, M., Motta, De Castro, E., Pérez, L.M., Jarrahbashi, D., *et al.*, Multifunctionality through Embedding Patterned Nanostructures in High-Performance Composites. *Adv. Mater.*, 35, 32, 1–10, 2023 Aug 29, https://onlinelibrary.wiley.com/doi/10.1002/adma.202300948.
234. Augustine, A., Qian, J., Faraone, T., Kolagatla, S., Prochukhan, N., Morris, M.A., *et al.*, Direct Laser Writing of Silica Nanoparticle Nanocomposites: Probing Mechanical Reinforcement and Understanding Structural Color from Design Parameters. *Small*, 20, 30, 38441362, 2024 Jul 5, https://onlinelibrary.wiley.com/doi/10.1002/smll.202310058.
235. Zhang, Y., Dai, S., Yin, Z., Yan, W., Li, Q., Yuan, H., *et al.*, Integration fabrication of polyimide composite films for aerospace applications. *SmartMat*, 5, 3, 1–13, 2023 Jun 21, Available from: https://onlinelibrary.wiley.com/doi/10.1002/smm2.1225.
236. Wu, C., Xu, F., Wang, H., Liu, H., Yan, F., Ma, C., Manufacturing Technologies of Polymer Composites—A Review. *Polymers*, 15, 3, 712, 2023 Jan 31, https://www.mdpi.com/2073-4360/15/3/712.
237. Omiogbemi, I.M.B., Yawas, D.S., Das, A., Afolayan, M.O., Dauda, E.T., Kumar, R., Gorja, S.R.S.R., Chowdhury, S.G. Mechanical properties and corrosion behaviour of duplex stainless steel weldment using novel electrodes. *Sci. Rep.*, 12, 22405, 2022. https://doi.org/10.1038/s41598-022-26974-6
238. Xiang, P., Zhang, H., Zhang, Q., Yu, Z., Wang, Y., Li, J., *et al.*, Mild Fabrication of Polymeric Porous/Nonporous Micro-Composite Structures *via* Enhancing the Photothermal Conversion of Ultrafast Laser. *Adv. Funct. Mater.*, 2024 Sep 9, https://onlinelibrary.wiley.com/doi/10.1002/adfm.202410304.
239. Parveez, B., Kittur, M.I., Badruddin, I.A., Kamangar, S., Hussien, M., Umarfarooq, M.A., Scientific Advancements in Composite Materials for Aircraft Applications: A Review. *Polymers*, 14, 22, 5007, 2022 Nov 18, https://www.mdpi.com/2073-4360/14/22/5007.
240. Ince, J.C., Peerzada, M., Mathews, L.D., Pai, A.R., Al-qatatsheh, A., Abbasi, S., *et al.*, Overview of emerging hybrid and composite materials for space applications. *Adv. Compos. Hybrid Mater.* [Internet], 6, 4, 130, 2023, Available from: https://doi.org/10.1007/s42114-023-00678-5.
241. Zheng, X., Zhang, H., Liu, Z., Jiang, R., Zhou, X., Functional composite electromagnetic shielding materials for aerospace, electronics and wearable fields. *Mater. Today Commun.*, 33, 104498, 2022. https://doi.org/10.1016/j.mtcomm.2022.104498
242. Piggott, M.R., The effect of the interface/interphase on fiber composite properties. *Polym. Compos.* [Internet], 8, 5, 291–7, 1987, Available from: https://4spepublications.onlinelibrary.wiley.com/doi/10.1002/pc.750080503.
243. Zhang, Z., Bellisario, D., Quadrini, F., Jestin, S., Ravanelli, F., Castello, M., *et al.*, Nanoindentation of Multifunctional Smart Composites. *Polymers*

(Basel) [Internet], 14, 14, 2945, 2022, Available from: https://www.mdpi.com/2073-4360/14/14/2945.

244. Islam, M.H., Afroj, S., Uddin, M.A., Andreeva, D.V., Novoselov, K.S., Karim, N., Graphene and CNT-Based Smart Fiber-Reinforced Composites: A Review. *Adv. Funct. Mater.* [Internet], 32, 40, 2205723, 2022. Available from: https://onlinelibrary.wiley.com/doi/10.1002/adfm.202205723
245. Dulal, M., Islam, M.R., Maiti, S., Islam, M.H., Ali, I., Abdelkader, A.M., *et al.*, Smart and Multifunctional Fiber-Reinforced Composites of 2D Heterostructure-Based Textiles. *Adv. Funct. Mater.* [Internet], 33, 40, 1–13, 2023, Available from: https://onlinelibrary.wiley.com/doi/10.1002/adfm.202305901.
246. Zhang, X., Chen, Y., Hu, J., Recent advances in the development of aerospace materials. *Prog. Aerosp. Sci.* [Internet], 97, February, 22–34, 2018, Available from: https://linkinghub.elsevier.com/retrieve/pii/S0376042117301483.
247. Liu, L., Guo, X., Liu, W., Lee, C., Recent Progress in the Energy Harvesting Technology—From Self-Powered Sensors to Self-Sustained IoT, and New Applications. *Nanomaterials* [Internet], 11, 11, 2975, 2021, Available from: https://www.mdpi.com/2079-4991/11/11/2975.
248. Maiti, S., Islam, M.R., Uddin, M.A., Afroj, S., Eichhorn, S.J., Karim, N., Sustainable Fiber-Reinforced Composites: A Review. *Adv. Sustain. Syst. [Internet]*, 6, 11, 2200258, 2022. Available from: https://onlinelibrary.wiley.com/doi/10.1002/adsu.202200258
249. Sanyal, S., Park, S., Chelliah, R., Yeon, S.J., Barathikannan, K., Vijayalakshmi, S., *et al.*, Emerging Trends in Smart Self-Healing Coatings: A Focus on Micro/Nanocontainer Technologies for Enhanced Corrosion Protection. *Coatings* [Internet], 14, 3, 324, 2024, Available from: https://www.mdpi.com/2079-6412/14/3/324.
250. Choi, W., Advincula, R.C., Wu, H.F., Jiang, Y., Artificial intelligence and machine learning in the design and additive manufacturing of responsive composites. *MRS Commun.* [Internet], 13, 5, 714–24, 2023, Available from: https://doi.org/10.1557/s43579-023-00473-9.

7

Behavioral Study of Tribological Coating of Smart Functional Composites for High Wear Applications

M. Sasi Kumar[1]*, N. Venkatesh[2], M. Makesh Kumar[3], S.L. Pradeep Kumar[1], D. Santhosh Kumar[1] and B. Deeban[1]

[1]Department of Aeronautical Engineering, KIT-Kalaignarkarunanidhi Institute of Technology, Tamil Nadu, India
[2]Department of Aerospace Engineering, SNS College of Technology, Tamil Nadu, India
[3]Department of Mechanical Engineering, KPR Institute of Engineering and Technology, Tamil Nadu, India

Abstract
Tribological coatings in smart functional composites are a primary study area contributing to the design of materials put to wear in several sectors. This chapter highlights the characteristics of the behavior of these coatings in terms of friction and wear resistance while simultaneously improving mechanical performance under extreme operating conditions. Moreover, the article embraces the concept of integrating intelligent materials into advanced composites that assimilate environmental stimuli such as temperature, pressure, and loading cycles. This work combines experimental analyses with theoretical modeling to elaborate on the mechanism within the microstructure, self-healing mechanisms, and the quality of the role that nanoscale reinforcement plays in constituting the proper ultimate improvement within the platform of tribological performance. The findings provide an insight into how these novel coatings can increase component life, reduce their maintenance needs, and provide sustainable engineering solutions for the high-demand sectors of aerospace, automotive, and energy.

**Corresponding author*: sasikumar.mak@gmail.com

Sandip Kunar, Pranav Charkha, Santosh Jaju and Harish Tiwari (eds.) Functional Composites: Role in Modern Engineering, (169–188) © 2025 Scrivener Publishing LLC

Keywords: Tribological coatings, smart functional composites, high-wear applications, adaptive materials, microstructural evolution, nanoscale reinforcements

7.1 Introduction

Composite materials join the inherent properties of the matrix with the improved or advanced functionalities added by dispersed particles. The matrix contributes with repairable properties, baseline mechanical functions, and mechanical integrity, while the dispersed particles add to the betterment or, at least, an additional function to the system. These enhancements are factors of particle content, uniformity of dispersion, material composition, and geological shape. Interactions within and outside the matrix and dispersed particles bring a new application class of materials that offer enhanced performance and flexibility [1]. Composite materials are formed to merge together the intrinsic properties of the matrix with the enhanced or advanced functionalities added by dispersed particles. The matrix provides itself with repairable properties, basic mechanical functions, and mechanical integrity, while the dispersed particles may add some enhancement and, let us not forget, at least one additional function to the system. These enhancements depend upon the particle content, uniformity of dispersion, composition of the material, and geological shape. Interaction within and out of the matrix and dispersed particles bring forth new materials that operate with greater performance and flexibility for specific applications [2]. Composite materials are designed to join the inherent properties of the matrix with the enhanced or advanced functionalities contributed by dispersed particles. While the matrix contributes repairable properties, baseline mechanical functions, and mechanical integrity, the dispersed particles add to these properties or are, at least, an additional functionalization of that system. All of these enhancements depend on particle content, uniformity of dispersion, material composition, and geometric shape. Interaction within and outside the matrix and dispersed particles brings about new materials for specific applications with enhanced performance and flexibility [3]. Nanomaterials are highly effective in antibacterial applications because of their unique physicochemical properties, which are governed by their massive surface area and size- or shape-dependent features. Because of these properties, nanoparticles can closely interact with bacterial cells to alter their structure and function. Mechanisms include the generation of reactive oxygen species (ROS), direct physical harm to cell membranes, and interference with microbial metabolic pathways, which are thought to be responsible for their antibacterial activity. Nanoparticles' versatility allows them to be applied to a wide range of platforms, including medical equipment, coatings, and drug delivery systems, offering innovative approaches to combat microbial infections [4]. Electrochemical tests correlate with the corrosion behavior of materials, and due to this, they find relevance in corrosion assessment. Of all the electrochemical methods, open circuit potential (OCP) is a widely used technique because it determines the equilibrium potential of any given material immersed in a specific electrolyte. OCP assesses how spontaneous the reaction is and combines the anodic and cathodic reactions into one equilibrium. It is used as a qualitative measure of the tendency of a material to corrode and thus helps in determining the corrosion resistance of the material and the integrity of its passive layer. It is possible to understand how various materials behave under corrosive environments and their degree of corrosion by monitoring OCP for an extended period [5]. The study examines the technological methods used to incorporate particles into coatings and how these methods affect the structure of the coating. It explores various factors that make it easier to add particles in order to enhance the properties of coatings such as hardness, friction, wear, abrasion, and corrosion resistance. Additionally, the discussion proposes that coatings containing particles offer improved resistance to high-temperature oxidation.

7.2 Principle of Tribology and Mechanisms of Wear

Tribology is the study of how moving surfaces interact with each other, especially in terms of friction, wear, and lubrication. Basic principles of trigonometry are important in understanding how materials behave under mechanical stress, factors such as surface roughness, lubricity, and material properties. By controlling biological stereo-phenomena, this affects the performance and longevity of the components.

7.2.1 Wear Types

Wear occurs when two surfaces come into contact and material is lost due to friction and other mechanical action. There are several key types of wear as shown in Figure 7.1.

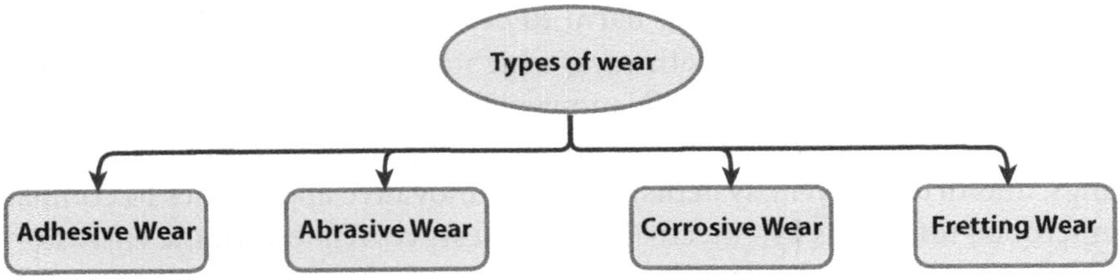

Figure 7.1 Types of wear.

Adhesive wear: The result of material transfer between two sliding surfaces due to height.
Abrasive wear: Caused by the action of hard particles or rough surfaces that scratch or gouge the material, causing the material to fall off.
Corrosive wear: Occurs when the environment is corrosive, such as moisture or chemicals; reacts with materials and accelerates wear from a chemical reaction.
Fretting wear: Microscopic movement between surfaces. This causes repeated friction and material erosion.

7.2.2 Significance of Coating in Enhancing Wear Resistance

Coatings are important for improving wear resistance. They do this by adding protective layers that lower friction, wear, and damage to materials. Tribological coatings like hard ceramics, thin films, and lubricating layers serve as shields, safeguarding surfaces from abrasive, adhesive, and corrosive forces. These coatings enhance surface hardness, decrease friction coefficients, and improve material adhesion, resulting in longer component lifespans, lower maintenance expenses, and better performance in areas prone to heavy wear.

7.3 Intelligent Multifunctional Composites

Smart functional composites are advanced materials designed to exhibit adaptive behavior in response to environmental stimuli such as temperature, pressure, and loading conditions. These composites combine multiple phases like matrix and reinforcements to enhance performance, including improved mechanical properties, wear resistance, and self-healing capabilities.

7.3.1 Categories of Functional Composites

Functional composites can be classified in terms of stimulus responsiveness. Normally, these composites require external energy changes, such as heat, to switch properties, but in the case of self-healing composites, the material itself will provide the triggering agent by opening up microcapsules that carry healing agents when the material suffers from damage. In addition, nanoparticle-reinforced composites, such as carbon nanotubes or graphene, are used to obtain better wear resistance and mechanical performance.

7.3.2 Applications in High-Wear Environments

They have huge utility in high-wear environments for the aerospace, automotive, and strength sectors. In aerospace, they decorate the power of key parts that have to perform in severe environments inclusive of accelerated temperatures and varying hundreds. In automotive applications, these composites offer friction and mechanical strain packages with improved wear performance and decreased upkeep expenses. Energy: For the energy enterprise, harsh environments, together with corrosion or excessive temperatures, are where they are used on the way to deliver longer component life and better reliability.

7.4 Tribological Coatings

Electrolytes of electrodeposited Co–Ni–Al_2O_3 composite coatings were made of 200–300 gm/L Co $(NH_4)_2SO_4{\cdot}4H_2O$ and 300–3,500 gm $Ni(NH_4)$.$(NH_4)_2SO_4{\cdot}4H_2O$ with 20 ml/L CH_3NO_2. Additive into the electrolyte, α-Al_2O_3 powder is added between 20 and 140, respectively. That has put the deposition of composite Co–Ni coatings, bearing the Co–Ni metal matrix with [α-Al_2O_3] reinforcement, which will lead to high performance, significantly improving the mechanical and wear properties [6]. With a total of 240 g/L of nickel sulphate ($NiSO_4$), 45 g/L of nickel chloride ($NiCl_2$), 30 g/L of boric acid (H_3BO_3), and 10 g/l of sodium dodecyl sulphate ($CH_3(CH_2)_{11}OSO_3Na$), well-purified Ni deposits were derived. This plating bath was prepared at a pH of 4.5. Using nickel sulphate, nickel ions were supplied to the bath, while nickel chloride was useful for conductivity and enhanced the quality of the coating. Buffered pH with boric acid maintained a constant electrochemical environment, and surfactant levels of sodium dodecyl sulphate were found to increase the adhesion and surface smoothness [7]. The coating was applied *via* immersing the substrate in answer at a rate of 15 cm/min using the usage of a computer-managed linear placing. The immersion setup changed into a maintained temperature and humidity-managed environment (25°C and 40% humidity), after which the ensuing gel film was heated at 500°C for a 30-min procedure. Repeat four instances to make certain a fair and sturdy coating [8].

The material describes how a flake-lined coating reduces the penetration of hydrochloric acid and other solvents. This reduced permeability is a key factor in reducing corrosion rates. By limiting the spread of corrosive materials, the aligned sections enhance the safety potential of the coating. Improve sturdiness and corrosion resistance in vital applications [9]. SiC (silicon carbide) and Si_3N_4 (silicon nitride) debris are blended in an electrolyte tub to attain a NiP composite coating. These particles are stored using a magnetic stirrer and Co bubbles. The mixture of SiC and Si_3N_4 debris is aimed to decorate the ternary properties of NiP composite coatings, such as wear resistance and overall mechanical performance. The efficient distribution of these particles ensures uniform coating properties. This makes it ideal for high wear and tear applications in aerospace, automotive, energy, and other industries [10]. The behavior of nanoscale aluminum oxide C and titanium oxide P25 particles provided by Degussa Hüls AG in aqueous electrolyte and particle size was investigated. and zeta potential measurements were performed using a Malvern Zetasizer 3000HS. The aim of the study was to understand the stability, diffusion, and surface properties of these nanoparticles in aqueous solution. The results indicate variations in particle size and zeta potential depending on the electrolyte composition. This famous essential insight into balance. Spread conduct and the viable programs of those nanoparticles in areas consisting of coatings and catalysis [11]. Studying the consequences of chemical composition on the overall performance of coatings in aluminum alloy structures is critical due to their numerous homes and packages. Factors influencing composition, such as adhesion, wear resistance, and protection against corrosion of the coating. Understanding these effects helps optimize coating formulations for better performance in various industrial applications [12]. TiO_2 particles have a narrow size range of 2.4–3 μm. Si_3N_4 particles show a wide size distribution: 20% in the range 8–10 μm, 40% in 6–8 μm, 10% in 3 to 4 μm, 20% in 2 to 3 μm, and 10% in 16–28 μm. The size of CeO_2 particles is 40% at 8–10 μm, 30% at 6–8 μm, 20% at 4–6 μm, and 10% at 13–17 μm [13].

A commercially available airborne wire arc gun system can be used to drill SHS717 core wire and develop nanocomposite microstructures. This process allows for the production of highly durable coatings with improved wear resistance and structural integrity. This technique takes advantage of rapid hardening and micro structuring to improve performance in high-wear applications [14, 15]. When ultrasound is applied to a liquid medium, the propagation of mechanical waves through a liquid result in the formation of sound cavities. These waves alternately pass through compression and purification cycles. This causes a change in pressure that leads to the formation of small bubbles. The collapse of these bubbles leads to an intense release of energy. Promote effects such as increased mixing particle distribution and cleaning the surface [16]. Use of electroless coating A Ni-based composite is coated on a 15 mm diameter test wheel specimen made of HB86 hardness low carbon steel. The objective of the process is to increase wear resistance and mechanical performance. These coated specimens are designed to study stereo biological behavior and improve durability in high-wear conditions [17]. The addition of sterols has been shown to improve adhesion to fruit and vegetable surfaces and coating films. This is because of the good agreement between the hydrophobic properties of the surface. The presence of sterols reduces the surface energy difference and increases the adhesion strength. This results in a more efficient coating.

7.5 Techniques for the Design and Fabrication of Materials

Titanium was coated using a dip-coating technique at a speed of 15 cm/min and sintered at 500°C for 30 min. The dip-coating process was repeated four times for each sample to ensure deposition of the coating. Regularly, the objective of this method is to enhance the triangular properties of the coating for high-wear applications [18]. Experimental techniques for characterizing coatings. Integrate electrochemical impedance spectroscopy (EIS) into long-term corrosion porosity studies under controlled flow conditions. SEM imaging should be supplemented with cross-sectional SEM analysis of the coating substrate and particle–matrix interface, while SAED-coupled TEM provides important phase composition information, atomic force microscopy (AFM) also enhances hardness assessment, micro-levels, and a wide range of stereo biology tests. Including dry, wet, and solution environments is essential to better understanding wear mechanisms in composite materials. Especially materials with nanometer-sized inclusions [19], nanostructured composite coating has better wear and corrosion resistance. This makes it suitable for modern applications that require improved surface properties. These coatings provide excellent mechanical performance. Especially in harsh environments, the complexity of the nanoscale structure increases durability. This makes it suitable for specialized applications with high wear and corrosion risk [20].

The nanostructures were investigated using high-resolution TEM (HR-TEM, JEOL 4000 EX/II at 400 kV) and analytical TEM (JEOL 2010F-FEG at 200 kV) in bright field (BF) to characterize the planar cross-sectional samples. Of the coating., high-resolution (HR) techniques such as selected field electron diffraction (SAED), energy filtering (EF), and imaging are used. These methods allow detailed characterization of nanostructures. It helps to understand their microstructural properties [21]. Electrolytic hydrogen evolution in composite coatings using steady-state polarization and electrolytic impedance spectroscopy (EIS) methods using the experimental method. The corrosion resistance and electrochemical behavior of the coating under hydrogen evolution conditions were analyzed. The results highlight the effectiveness of the coating in reducing hydrogen evolution. Better corrosion resistance and demonstrates its potential for use in environments involving hydrogen [22]. This is to keep the particles suspended and prevent overheating. An electroless nickel coating was then applied to 2,000 mL of bath solution using a magnetic stirrer. All deposits were plated at 87°C and pH 4.6 to ensure a uniform phosphorus content. The objective of this controlled condition is to achieve uniform properties throughout the installed piece [23, 24]. For SEM observation, metallurgical cross-sections were ground and polished in such a way that they did not contaminate. Avoid SiC sandpaper to prevent contamination, and EDX analysis was used on these polished sections to determine the amount of SiC particles co-located in the coating. This helps ensure an accurate assessment of coating composition without interference from corrosive pollution [25]. The graphite composition has a sliding pressure range of 0.001–0.002 MPa. The fibrous material weighing 20 ± 1 g is produced homogeneously in terms of moisture, impurities, and main properties in accordance with regulatory and technical standards [26].

7.6 Development of Performance

It is worth noting that at the highest cross speed you will notice unique behavior. Especially for the 6061 Al alloy, this indicates that the alloy performance and response of the ternary coating may vary significantly with increasing speed. The findings highlight the importance of understanding how different processing parameters affect the wear behavior of materials such as 6061 Al alloy, which may influence coating development and optimization. For applications with high wear and tear [27], electrolyte solution is used for composite coating deposition. HA/chitosan was prepared by combining chitosan and calcium phosphate solutions. These composite coatings are prepared by combining the natural polymer chitosan with calcium phosphate. It is known for its biocompatibility and role in promoting bone-like structure. A uniform and cohesive coating of the composite HA/chitosan can do this, which improves the mechanical and biological properties as well as surface properties. This approach promises the development of coatings with improved wear resistance, adhesion, and performance in biomedical and high-wear applications [28]. Zeta potential measurements were performed using a Malvern laser zeta meter (Zetasizer 2000). Ground samples weighing 0.02 g were dispersed in 50 mL of mixed twist water, with 10–2 M NaCl solution added to control the strength, ionic, and adjust the pH of the suspension to the desired value using NaOH solution. The sample was shaken for 30 min to ensure complete diffusion. After that, the potential zeta was measured. This approach provides important insights into the surface charge and stability of the sample. This is important for understanding the behavior of tribological coatings in aqueous environments [29]. Results from this test are expressed as weight loss per unit coating area (mg/mm^2), allowing coating performance to be compared regardless of specimen size. This method normalizes the data, making it applicable to coatings on materials of different sizes. By expressing the results in terms of weight reduction per area. This allows for a more efficient assessment of the ternary behavior and durability of coatings. It facilitates direct performance comparisons between various coating systems [30]. An austenitic stainless steel sheet is used as the anode, while the prepared samples act as cathodes in the experimental setup. This setup facilitates the study of ternary coatings, where the anode is the source of oxidation, and the cathode is responsible for the absorption and reaction of the coating. This is important for analyzing the performance of tribological coatings used in smart functional composites [31]. The diffusion layer method assumes the absence of convection and intermediate reactions where the adsorption of Ni and Co ions is partially reduced. This process relies on the creation of a diffusion layer. Where ions migrate to the surface. This is mainly driven by diffusion rather than convection forces occurring in systems that control diffusion. This approach reduces the effect of external convection forces. This ensures a more controlled and uniform coating deposition in specific applications [32]. The frame with $10 \times 10 \times 5\text{-mm}^3$ perforations was compression tested using the Instron model 4505 (UK) in ambient conditions. A 1-N load cell was used to apply the load at a crosshead speed of 2 mm/min. The resulting stress–strain curves were analyzed to investigate the mechanical properties of the frame, such as compressive strength and elasticity parameters. This test method provides insights into the mechanical behavior of porous scaffolds. This is necessary for evaluating suitability for structural or biomedical applications [33]. MWCNT-xylene/PDMS coating displays the smallest peak-to-peak pressure amplitude. This indicates good pressure control performance. However, it has the largest bandwidth (39.8 MHz) and highest frequency (28.5 MHz) compared to other coatings. This shows that although these coatings effectively reduce the peak pressure amplitude, they can work in a wider frequency range. This makes it ideal for applications that require wide frequency coverage and improved dynamic performance [34]. The overall coating profile showed a consistent trend. It has a higher initial value compared to polyethylene (PE) coatings. As the depth increases beyond 150 nm, the coatings converge until the values are almost equal. This indicates that although the properties of the composite coating may be improved at shallow depths, its performance is normal at greater depths. This indicates comparable behavior to polyethylene coatings over a wide range [35]. HA nanoparticles show a higher deposition

Table 7.1 Performances of materials and its observations.

Material	Mechanical conditions	Key findings	Reference
Ag/HAP/Gr	183.0 ± 21.9 GPa	Increased elastic modulus	[37]
MoS_2–Cr and MoS_2–Ti	7.5 GPa and 8.4 GPa	Increased relative humidity	[38]
Polyethylene-free nickel deposits	7.5–9 A/dm^2 at pH 4	Increase wear resistance	[39]
$ZnSO_4$ $6H_2O$, 80 g/ Na_2SO_4	Stirrer at 1,000 rpm	Increased surface hardness	[40]
NiP/SiC	800°C	Improved mechanical properties	[41]
Ni–AlN	100–650°C	High resistance	[42]
Nickel–aluminum	40–50 mm	Improved thickness	[43]

rate than GO/HA particle aggregates, resulting in higher coating thicknesses. HA coatings show less corrosion resistance behavior than GO/HA composite coatings [36]. Table 7.1 presents the observation of the different materials and its conditions.

7.7 Obstacles and Constraints

7.7.1 Material Compatibility Challenges

Several rather important limitations and challenges affect tribological coatings on smart functional composites. One major challenge is material compatibility, where adhesion, endurance, and overall performance of coatings must still be optimized when working with multiple substrates. Poor bonding, wear resistance, and functionality can result from differences in the thermal, mechanical, and chemical properties of coatings and composites. Generating coatings likely to appropriately complement such smart composites in a range of functioning conditions is yet a multifaceted difficulty.

7.7.2 Considerations Regarding Expenses and the Ability to Scale

Another major challenge is that of cost and scalability. The production of advanced tribological coatings, in general, involves high material costs and processing costs. This makes them less viable for widespread industrial applications or large-scale applications. Increasing the production volume while maintaining the coating quality brings forward several concerns related to the financial aspects that directly affect the economic feasibility of these technologies.

7.7.3 Constraints Related to Environmental and Operational Factors

Lastly, there are environmental and operational challenges that need to be hurdled. In terms of envwironmental sustainability, tribological coatings must operate in extremes of temperature, in corrosive environments, and under fluctuating loads. This requires that such coatings should not only be deposited in an environmentally sustainable process but also withstand being operated under these conditions without causing harmful emissions. To ensure robust and reliable tribological performance of the coatings, it is important to consider any effect the environment may have on their operation over long periods of time into the future. And lastly, increased awareness in the industry concerning minimization of use of critical raw materials and resources due to both their increasing costs and potential geopolitical constraints, quantities where there has been concern expressed regarding human health effects or environmental impact, substances subject to export restrictions when regulations become more stringent, non-sustainability leading to either declines in quality or quantity, thus hindering some engineering applications. These hurdles can only be overcome by investing further efforts into investigating new coating materials, developing new coating manufacturing processes, and adopting/coating existing or new materials using green, sustainable approaches.

7.8 Emerging Trends and Research Pathways

The most important assignment in growing nanocomposite coatings is to increase their shelf life and manage their houses more. It is crucial to understand that they have an impact on factors that include thermal, mechanical, and chemical balance on coating performance. Addressing this undertaking permits the development of environmentally friendly coatings on a larger scale and at a lower price. This ensures durability and economic feasibility. Improved control of these properties will affect the long-term durability and performance of nanocomposite coatings [44]. The main challenge of integrating graphene into a nickel matrix composite coating is to ensure uniform distribution and strong adhesion between the graphene particles and the nickel substrate. Future realizations to optimize electrochemical reduction strategies to enhance graphene dispersion, reduce defects, and improve coating adhesion and mechanical facilitation are also provided. It is much more important to test the overall performance of these coatings over long periods of time in harsh environments such as extreme temperatures and corrosive atmospheres. Finding process scalability and reducing production costs will be important for industrial applications [45]. Optimal dispersion of cellulose nanocrystals (CNC) and microcrystals (CMC) in methacrylic siloxane resin cured with ultraviolet light, to enhance beneficial properties such as thermal stability and mechanical performance without affecting the flexibility of the matrix. Future visions will likewise recognize improvements in CNC dispersion methods, supporting integration within the matrix. Long-term sturdiness below special environmental conditions is being investigated. Additionally, opportunity resin formulations are being explored and huge-scale application testing is being conducted. This is especially true for wood coatings. In broad industrial use, moisture resistance can also improve performance in moisture. The SiC content in Ni–W/SiC nanocomposite coatings is used to achieve the desired mechanical properties without affecting the integrity of the coating. Future work will focus on finding alternative ways to control SiC particle agglomeration, especially improving the thickness of the pulse diffusion layer to optimize particle size matching. In addition, finding new deposition techniques or additives that can reduce the grain size refinement effect above 5 vol.% is increasingly important to increase the mechanical properties. With the development of methods for balancing stiffness and modulus while maintaining durability and wear resistance, this will make high-wear applications more reliable [46].

Uniform distribution of fine SiC nanoparticles within the nickel matrix is guaranteed. This is especially true under different deposition conditions such as direct pulse current. This makes it difficult to achieve an optimal particle–matrix interaction while maintaining the integrity and mechanical properties of the coating. Future research will focus on improving deposition techniques to locate stepped forward alloys or coatings and to assess their lengthy-time period performance beneath harsh environmental conditions. In addition, the electrochemical behavior of these coatings might be, in addition, investigated. It is critical to boom sturdiness and practicality, particularly in corrosive environments in the aerospace, car, and electricity industries [47] by understanding the precise mechanism by which TiO_2 debris interact with the coating shape and affect the HA grain growth during re-solidification. Further research is needed to determine the lengthy time period stability and sturdiness of these chemical reactions under special environmental operating situations. Future work will focus on optimizing the microstructure development of the coating, increasing mechanical properties, and evaluation of biological behavior in real use. There may be further improvements [48] in current density optimization for specific *in situ* systems to enhance the efficiency and properties of the deposited material, such as mechanical strength or corrosion resistance. Challenges include modeling the complex interaction of current densities and precise adsorption and deposition kinetics to ensure scalability for industrial applications. The effect of process parameters on adhesion can be examined. In terms of microstructure and long-term stability, developing prediction models to optimize loose and strong drying will be important [49], with specific microstructural analysis. Thus, the presence and distribution of Cu nanocrystals in the Ta-rich Cu–Ta matrix may be accurately quantified. Advanced characterization strategies inclusive of high-resolution electron microscopy and X-ray diffraction are crucial. In addition, mechanical checks such as tensile and effect tests are important. It could be essential to assess flexibility upgrades. Challenges include ensuring a uniform distribution of the nanocrystals, dealing with oxidation or surface reactions which can arise, and know-how of the lengthy-term stability of coatings underneath diverse situations [50]. CNT fabric is employed for maximum mechanical properties without compromising biocompatibility. It is vital to research the long-term balance and bioactivity of the CNT-reinforced hydroxyapatite coating in physiological situations. In addition, the scalability and cost-effectiveness of the manufacturing procedure need to be explored. Two challenges consist of making sure that there is a uniform distribution of CNTs, preventing aggregation and handling possible poisonous risks. Future studies have to examine the medical efficacy and long-term outcomes of those coatings *in vivo* to verify their efficacy in bone repair [51].

The bottom of form optimizing the proportion of PTFE particles in composite coating to offer improved put-on resistance, affecting other homes such as mechanical electricity, by exploring the effect of various processing techniques, together with advanced deposition strategies. It additionally offers insights into reducing wear prices. Challenges include dealing with the impact of different environmental situations (e.g., temperature, humidity) on wear performance. This ensures that a regular coating is pleasant. Moreover, for business applications, the improvement of scalable and price-effective production techniques remains a key location of investigation [52]. Electronically deposited Ni–B/SiC micro- and nanocomposite coatings pay attention to optimizing the general shape to enhance mechanical, put-on, and corrosion resistance. At the same time, it reduces porosity. It is essential to develop a better understanding of the accumulation mechanism. Especially the role of SiC particle diffusion, future challenges include achieving uniform coating properties at the micro-nano scale by manipulating the interfacial bonds between Ni–B particles and the SiC matrix and scaling up the process for industrial applications in addition to the significant challenge of improving durability and performance in harsh environments [53].

7.9 Conclusion

The investigation of tribological coatings in smart functional composites is referenced as highly important for the future of high-wear applications in industries such as aerospace, automotive, energy, and more. The characteristics and behavior of these coatings, including their friction-reducing capabilities and improved resistance to wear and mechanical performance under extreme conditions of high temperature, pressure, and variable loads, are also described in this chapter. The investigation includes the incorporation of smart materials and smart composites with various adaptive responses, analysis of microstructural evolution, self-repair mechanisms, and nanoscale reinforcements that can improve tribological performance. The guidance obtained through these approaches is essential for optimizing coating performance and durability during field applications. The findings support engineering solutions to achieve sustainability. By reducing the maintenance costs and extending the life of the component, this would be beneficial. The acceptance of coatings will positively affect the application of this technology, leading to innovation and dependability in strategic industries. Compatibility, cost, and environmental compatibility are provided specifically to ensure that their successful implementation is not hampered.

References

1. Lelevic, A. and Walsh, F.C., Electrodeposition of Ni–P composite coatings: A review. *Surf. Coat. Technol.*, 378, 124803, 2019. doi: 10.1016/j.surfcoat.2019.07.027.
2. Walsh, F.C., Wang, S., Zhou, N., The Electrodeposition of Composite Coatings: Diversity, Applications and Challenges. *Curr. Opin. Electrochem.* 20, 8–19, 2020.
3. Agarwala, R.C. and Agarwala, V., Electroless alloy/composite coatings: A review. *Sadhana,* 28, 475–493, 2003.
4. Guo, L., Yuan, W., Lu, Z., Li, C.M., Polymer/nanosilver composite coatings for antibacterial applications. *Colloids Surf. A Physicochem. Eng. Asp.*, 439, 69–83, Dec. 2013, doi: 10.1016/j.colsurfa.2012.12.029.
5. Huang, Y.S., Zeng, X.T., Hu, X.F., Liu, F.M., Corrosion resistance properties of electroless nickel composite coatings. *Electrochim. Acta*, 49, 25, 4313–4319, Oct. 2004, doi: 10.1016/j.electacta.2004.04.023.
6. Wu, G., Li, N., Zhou, D., Mitsuo, K., Electrodeposited Co-Ni-Al O composite coatings 2 3. *Surf. Coat. Technol.*, 176, 157–164, 2004, doi: 10.1016/S0257-8972Ž03.00739-4.
7. Lekka, M., Kouloumbi, N., Gajo, M., Bonora, P.L., Corrosion and wear resistant electrodeposited composite coatings. *Electrochim. Acta*, 50, 23 SPEC. ISS., 4551–4556, Aug. 2005, doi: 10.1016/j.electacta.2004.11.067.
8. Milella, E., Cosentino, F., Licciulli, A., Massaro, C., Preparation and characterisation of titania/hydroxyapatite composite coatings obtained by sol–gel process. *Biomaterials*, 22, 11, 1425–1431, 2001.

9. Yang, C., Smyrl, W.H., Cussler, E.L., Flake alignment in composite coatings. *J. Membr. Sci.*, 231, 1–2, 1–12, Mar. 2004, doi: 10.1016/j.memsci.2003.09.022.
10. Sarret, M., Müller, C., Amell, A., Electroless NiP micro- and nano-composite coatings. *Surf. Coat. Technol.*, 201, 1–2, 389–395, Sep. 2006, doi: 10.1016/j.surfcoat.2005.11.127.
11. Erler, F., Jakob, C., Romanus, H., Spiess, L., Wielage, B., Lampke, T., Steinhäuser, S., Interface behaviour in nickel composite coatings with nano-particles of oxidic ceramic. *Electrochim. Acta*, 48.20-22, 3063–3070, Sep. 2003, doi: 10.1016/S0013-4686(03)00380-3.
12. Gnedenkov, S.V., *et al.*, Formation and properties of composite coatings on aluminum alloys. *Russ. J. Inorg. Chem.*, 62, 1, 1–11, Jan. 2017, doi: 10.1134/S0036023617010065.
13. Balaraju, J.N., Sankara Narayanan, T.S.N., Seshadri, S.K., Structure and phase transformation behaviour of electroless Ni-P composite coatings. *Mater. Res. Bull.*, 41, 847–860, Apr. 2006, doi: 10.1016/j.materresbull.2005.09.024.
14. Branagan, D.J., Breitsameter, M., Meacham, B.E., Belashchenko, V., High Performance Nanoscale Composite Coatings For Boiler Applications. *J. Therm. Spray Technol.*, 14, 196–204, 2005.
15. Vaezi, M.R., Sadrnezhaad, S.K., Nikzad, L., Electrodeposition of Ni-SiC nano-composite coatings and evaluation of wear and corrosion resistance and electroplating characteristics. *Colloids Surf. A Physicochem. Eng. Asp.*, 315, 1–3, 176–182, Feb. 2008, doi: 10.1016/j.colsurfa.2007.07.027.
16. Tudela, I., Zhang, Y., Pal, M., Kerr, I., Cobley, A.J., Ultrasound-assisted electrodeposition of composite coatings with particles. *Surf. Coat. Technol.* 259, 363–373, 2014.
17. Wang, L.Y., *et al.*, Friction and wear behavior of electroless Ni-based CNT composite coatings. *Wear*, 254, 12, 1289–1293, 2003, doi: 10.1016/S0043-1648(03)00171-6.
18. Ramires, P.A., Romito, A., Cosentino, F., Milella, E., The influence of titania/hydroxyapatite composite coatings on *in vitro* osteoblasts behaviour. *Biomaterials*, 22, 12, 1467–1474, 2001.
19. Walsh, F.C. and Ponce De Leon, C., A review of the electrodeposition of metal matrix composite coatings by inclusion of particles in a metal layer: An established and diversifying technology. *Transactions of the IMF*, 92.2, 83–98, 2014, doi: 10.1179/0020296713Z.000000000161.
20. Benea, L., Bonora, P.L., Borello, A., Martelli, S., Wear corrosion properties of nano-structured SiC-nickel composite coatings obtained by electroplating. *Wear*, 249, 10–11, 995–1003, 2002.
21. Galvan, D., Pei, Y.T., De Hosson, J.T.M., Deformation and failure mechanism of nano-composite coatings under nano-indentation. *Surf. Coat. Technol.*, 200, 24, 6718–6726, Aug. 2006, doi: 10.1016/j.surfcoat.2005.10.010.
22. Kubisztal, J., Budniok, A., Lasia, A., Study of the hydrogen evolution reaction on nickel-based composite coatings containing molybdenum

powder. *Int. J. Hydrogen Energy*, 32, 9, 1211–1218, Jun. 2007, doi: 10.1016/j.ijhydene.2006.11.020.

23. Apachitei, I., Duszczyk, J., Katgerman, L., Overkamp, P.J.B., Electroless Ni-P composite coatings: the effect of heat treatment on the microhardness of substrate and coating. *Scripta Materialia,* 38.9, 1347–1353, 1998.
24. Ranganatha, S., Venkatesha, T.V., Vathsala, K., Development of electroless Ni-Zn-P/nano-TiO 2 composite coatings and their properties. *Appl. Surf. Sci.*, 256, 24, 7377–7383, Oct. 2010, doi: 10.1016/j.apsusc.2010.05.076.
25. Garcia, I., Fransaer, J., Celis, J.-P., Electrodeposition and sliding wear resistance of nickel composite coatings containing micron and submicron SiC particles. *Surf. Coat. Technol.*, 148.2-3, 171–178, 2001.
26. Ziyamukhamedova, U., Djumabaev, A., Urinov, B., Almatayev, T., Features of structural adaptability of polymer composite coatings. *E3S Web Conf.*, EDP Sciences, 264, 05011, Jun. 2021, doi: 10.1051/e3sconf/202126405011.
27. Kadolkar, P.B., Watkins, T.R., De Hosson, J.T.M., Kooi, B.J., Dahotre, N.B., State of residual stress in laser-deposited ceramic composite coatings on aluminum alloys. *Acta Mater.*, 55, 4, 1203–1214, Feb. 2007, doi: 10.1016/j.actamat.2006.07.049.
28. Redepenning, J., Venkataraman, G., Chen, J., & Stafford, N., Electrochemical preparation of chitosan/hydroxyapatite composite coatings on titanium substrates. *J. Biomed. Mater. Res. A,* 66.2, 411–416, 2003.
29. Aal, A.A., Zaki, Z.I., Hamid, Z.A., Novel composite coatings containing (TiC-Al_2O_3) powder. *Mater. Sci. Eng. A*, 447, 1–2, 87–94, Feb. 2007, doi: 10.1016/j.msea.2006.10.036.
30. Coan, T., Barroso, G.S., Motz, G., Bolzána, A., Machado, R.A.F., Preparation of PMMA/hBN composite coatings for metal surface protection. *Mater. Res.*, 16, 6, 1366–1372, Nov. 2013, doi: 10.1590/S1516-14392013005000140.
31. Parida, G., Chaira, D., Chopkar, M., Basu, A., Synthesis and characterization of Ni-TiO2 composite coatings by electro-co-deposition. *Surf. Coat. Technol.*, 205, 21–22, 4871–4879, Aug. 2011, doi: 10.1016/j.surfcoat.2011.04.102.
32. Karimzadeh, A., Aliofkhazraei, M., Walsh, F.C., A review of electrodeposited Ni-Co alloy and composite coatings: Microstructure, properties and applications. *Surf. Coat. Technol.,* 372, 463–498, Aug. 25, 2019. doi: 10.1016/j.surfcoat.2019.04.079.
33. Kim, H.W., Knowles, J.C., Kim, H.E., Hydroxyapatite/poly(ε-caprolactone) composite coatings on hydroxyapatite porous bone scaffold for drug delivery. *Biomaterials*, 25, 7–8, 1279–1287, 2004, doi: 10.1016/j.biomaterials.2003.07.003.
34. Noimark, S., *et al.*, Carbon-Nanotube–PDMS Composite Coatings on Optical Fibers for All-Optical Ultrasound Imaging. *Adv. Funct. Mater.*, 26, 46, 8390–8396, Dec. 2016, doi: 10.1002/adfm.201601337.
35. Chih, A., Ansón-Casaos, A., Puértolas, J.A., Frictional and mechanical behavior of graphene/UHMWPE composite coatings. *Tribol. Int.,* 116, 295–302, 2017.

36. Li, M., *et al.*, Graphene oxide/hydroxyapatite composite coatings fabricated by electrophoretic nanotechnology for biological applications. *Carbon,* Carbon N Y, 67, 185–197, 2014, doi: 10.1016/j.carbon.2013.09.080.
37. Janković, A., Eraković, S., Vukašinović-Sekulić, M., Mišković-Stanković, V., Park, S.J., Rhee, K.Y., Graphene-based antibacterial composite coatings electrodeposited on titanium for biomedical applications. *Prog. Org. Coat.*, 83, 1–10, Jun. 2015, doi: 10.1016/j.porgcoat.2015.01.019.
38. Ding, X.Z., Zeng, X.T., He, X.Y., Chen, Z., Tribological properties of Cr- and Ti-doped MoS2 composite coatings under different humidity atmosphere. *Surf. Coat. Technol.*, 205, 1, 224–231, Sep. 2010, doi: 10.1016/j.surfcoat.2010.06.041.
39. Hamid, Z.A. and Ghayad, I.M., Characteristics of electrodeposition of Ni-polyethylene composite coatings. *Mater. Lett.,* 53.4-5, 238–243, 2002, [Online]. Available: www.elsevier.com/locate/matlet.
40. Praveen, B.M., Venkatesha, T.V., Arthoba Naik, Y., Prashantha, K., Corrosion studies of carbon nanotubes-Zn composite coating. *Surf. Coat. Technol.*, 201, 12, 5836–5842, Mar. 2007, doi: 10.1016/j.surfcoat.2006.10.034.
41. Malfatti, C.F., *et al.*, NiP/SiC composite coatings: the effects of particles on the electrochemical behaviour. *Corr. Sci.,* 47.3, 567–580, 2005.
42. Aal, A.A., Bahgat, M., Radwan, M., Nanostructured Ni-AlN composite coatings. *Surf. Coat. Technol.*, 201, 6, 2910–2918, Dec. 2006, doi: 10.1016/j.surfcoat.2006.06.002.
43. Sharma, A.K., Aravindhan, S., Krishnamurthy, R., Microwave glazing of alumina-titania ceramic composite coatings, Materials Letters 50.5-6, 295–301, 2001. [Online]. Available: www.elsevier.comrlocatermatlet.
44. Riaz, U., Nwaoha, C., Ashraf, S.M., Recent advances in corrosion protective composite coatings based on conducting polymers and natural resource derived polymers. *Prog. Org. Coat.,* 77.4, 2014, 743–756, Apr. 2014, doi: 10.1016/j.porgcoat.2014.01.004.
45. Szeptycka, B., Gajewska-Midzialek, A., Babul, T., Electrodeposition and Corrosion Resistance of Ni-Graphene Composite Coatings. *J. Mater. Eng. Perform.,* 25, 3134–3138, Aug. 01, 2016, doi: 10.1007/s11665-016-2009-4.
46. Wasekar, N.P., Latha, S.M., Ramakrishna, M., Rao, D.S., Sundararajan, G., Pulsed electrodeposition and mechanical properties of Ni-W/SiC nanocomposite coatings. *Mater. Des.*, 112, 140–150, Dec. 2016, doi: 10.1016/j.matdes.2016.09.070.
47. Zanella, C., Lekka, M., Bonora, P.L., Influence of the particle size on the mechanical and electrochemical behaviour of micro- and nano-nickel matrix composite coatings. *J. Appl. Electrochem.*, 39, 1, 31–38, Jan. 2009, doi: 10.1007/s10800-008-9635-y.
48. Li, H., Khor, K.A., Cheang, P., Impact formation and microstructure characterization of thermal sprayed hydroxyapatite/titania composite coatings. *Biomaterials,* 24.6, 949–957, 2003.

49. Bahadormanesh, B. and Dolati, A., The kinetics of Ni-Co/SiC composite coatings electrodeposition. *J. Alloys Compd.*, 504, 2, 514–518, Aug. 2010, doi: 10.1016/j.jallcom.2010.05.154.
50. Bahrami, A., Onofre Carrasco, C.F., Cardona, A.D., Huminiuc, T., Polcar, T., Rodil, S.E., Mechanical properties and microstructural stability of CuTa/Cu composite coatings. *Surf. Coat. Technol.*, 364, 22–31, Apr. 2019, doi: 10.1016/j.surfcoat.2019.02.072.
51. Chen, Y., Zhang, Y.Q., Zhang, T.H., Gan, C.H., Zheng, C.Y., Yu, G., Carbon nanotube reinforced hydroxyapatite composite coatings produced through laser surface alloying. *Carbon,* Carbon N Y, 44, 1, 37–45, Jan. 2006, doi: 10.1016/j.carbon.2005.07.011.
52. Balaji, R., Pushpavanam, M., Kumar, K.Y., Subramanian, K., Electrodeposition of bronze-PTFE composite coatings and study on their tribological characteristics. *Surf. Coat. Technol.*, 201, 6, 3205–3211, Dec. 2006, doi: 10.1016/j.surfcoat.2006.06.039.
53. Salari Mehr, M., Akbari, A., Damerchi, E., Electrodeposited Ni-B/SiC micro- and nano-composite coatings: A comparative study. *J. Alloys Compd.*, 782, 477–487, Apr. 2019, doi: 10.1016/j.jallcom.2018.12.184.

8

Applications to Dynamic Behavior: Free Vibration Analysis of Functional Composites

Debarupam Gogoi*, Pramod Kumar Parida, Md. Irquam Alam, Mihir Kumar Pandit and Arun Kumar Pradhan

School of Mechanical Sciences, Indian Institute of Technology Bhubaneswar, Odisha, India

Abstract

Composite materials find their applications due to varying properties in different directions in a wide range of industries like aerospace, vehicle, medical, and many others. However, a delamination and stress concentrations at the interface between two materials are issues that composite materials must deal with owing to the distinction between the thermal coefficients of their layers. In 1984, the idea of functional materials was presented as a solution to this issue. The concept is of gradual variation of the mechanical, thermal, electrical, and magnetic properties of two or more materials in single or bi-directions to create synergistic effects, which is not possible by a single material paved the way for revolutionary solutions. The materials thus integrated exhibit specific functionalities such as lightweight structures, electromagnetic shielding, heat dissipation, and damping characteristics by improving mechanical strength, toughness, electrical and heat conductivity, energy dissipation, and corrosion resistance. The abstract refers to the recent composite materials with its advantages and drawbacks, the trend for advanced composite materials or functionally graded materials, different functionally gradient material (FGM) structures with their applications, and dynamic behavior of FGM plate structures.

***Keywords*:** FGM, FEM, free vibration, natural frequency

**Corresponding author*: a21me09006@iitbbs.ac.in

Sandip Kunar, Pranav Charkha, Santosh Jaju and Harish Tiwari (eds.) Functional Composites: Role in Modern Engineering, (189–214) © 2025 Scrivener Publishing LLC

8.1 Introduction

Composite materials, created by combining two or more distinct constituents, have significantly impacted modern industries on account of their ability to deliver tailored properties for specific applications. The recent years have seen significant advancements in these materials, especially in the form of materials with functional grades (FGMs) and advanced composite materials (ACMs). These materials have made possible to create strong and lightweight parts to be widely applicable in sectors like energy, biomedical, automotive, marine, and aerospace [1, 2]. ACMs, such as glass fiber and carbon fiber polymers, are famous for their significant strength-to-weight ratio, while FGM can provide a gradual transition of properties across a structure, which make them ideal with thermal and mechanical gradient in modern engineering applications [3, 4].

Both ACMs and FGMs pose significant benefits. The capacity of fiber-reinforced composites to provide structural strength while minimizing weight is highly admired, which is especially important in exploration of space and aviation sectors [5]. By progressively changing the material composition, FGMs, on the other way, provide remarkable mechanical and thermal performances that make them suitable for application in energy systems subjected to extreme conditions, biomedical implants, and thermal barrier coatings [6]. Additionally, nanocomposites, incorporating materials like graphene or carbon nanotubes, have further enhanced qualities of polymer composites, providing superior mechanical strength and thermal conductivity [7, 8]. Figure 8.1 shows a composite shell with constrained layers for damping purposes.

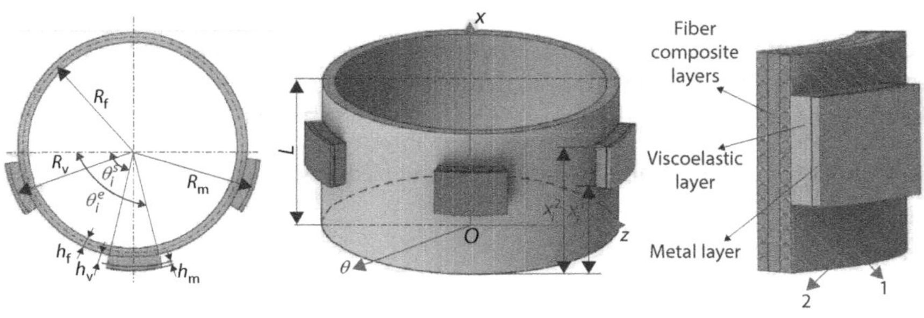

Figure 8.1 Schematic diagram showing the geometry of a composite shell with constrained layer damping [39]. (*Adapted with permission from Elsevier, license no. 5950180474412*).

Despite these benefits, challenges are associated with the manufacturing and use of these materials. High-performance ACMs, like carbon fiber composites, remain costly on account of complex fabrication processes, including high temperature curing with additive manufacturing techniques [9]. Recycling is another significant issue, as many composite materials are hard to disengage into their base parts, looking for environmental concerns [10]. Moreover, failure criterion like delamination, matrix cracking, and fiber pull-out may affect the structural integrity of composites, particularly in dynamic loading environments [1, 2]. Recent techniques aim to overcome these barriers while expanding the utility of composite materials. For instance, sustainable and biodegradable composites are gaining traction and leveraging natural fibers with bio-based polymers to reduce environmental impact [11]. Additive manufacturing has enabled the production of complex geometries and property gradients essential for FGMs, advancing their customization potential [3]. Furthermore, self-healing composites, which can autonomously repair damage, are being explored to enhance durability and reduce maintenance costs [8, 12].

This chapter explores the developments in FGMs, addressing their advantages, challenges, and future trends while providing insights into their expanding roles in modern engineering and technology. Additionally, functionally graded plate structure under dynamic behavior condition is also highlighted for illustration.

8.1.1 Advantages of FGM Over Traditional Composites

Innovative materials with functional gradation are made with gradual variation in composition across their thickness, which produces constantly varying material characteristics. Compared to traditional composites, in which different layers frequently cause sudden changes in characteristics, the FGM design has a number of benefits. Because of excellent properties with exceptional performance in difficult situations, FGMs have found their applicability in a range of industries involving aerospace, energy, biomedical, etc. The tailored property distribution inside FGMs made it possible for smooth transitions between two materials. Traditional composites frequently suffer from stress concentrations at the layer interfaces, which can cause early failure by delamination. By eliminating these abrupt changes along interfaces, FGM diminishes stress concentrations with a continuous gradation of characteristics [13]. For instance, FGMs are used as thermal barrier coatings in aerospace applications, where the material changes from a ceramic-rich portion that can withstand high heat to a metal-rich substrate that provides strength and toughness. FGMs can function dependably in worse environments due to its seamless transition, which improves structural stability and thermal load carrying capacity [14].

The superior thermal resistance of FGMs as compared to conventional composites, which frequently fail by mismatched coefficients of thermal expansion (CTE) between their interfaces, enables FGMs to have superior thermal resistance in high temperature settings, like those found in rocket engines or turbine blades. Thermal cracking or delamination may arise in conventional composite from imbalance caused by thermal coefficients. Conversely, FGMs are designed to show leisurely changes in CTE, which help to regulate heat stresses and provide durable stability [15]. In energy systems, where components must tolerate large temperature variations without degrading, this feature is especially advantageous. Hence, the weak points of conventional composites are diminished by the continuous nature of FGMs, which enhances structural integrity along with mechanical strength and durability [16]. For instance, FGMs are replacing biomedical implants that replicate the mechanical property gradient seen naturally in human bone. Because of this gradient, FGMs are highly applicable in orthopedics and the dental field, which enhance load distribution and lower the chance of implant rejection [17].

Multifunctionality is another significant benefit of FGMs. The adaptability of traditional composites is limited since they are usually optimized for a single feature, such as strength or heat resistance. On the other hand, FGMs can incorporate several functions into a single substance. For instance, an FGM may have a strong core to withstand mechanical shocks and a wear-resistant surface for durability. Applications where both surface durability and core toughness are crucial, such as protective armor and cutting tools, benefit greatly from this dual functionality [18].

Longer lifespans and greater durability make FGMs renowned. Owing to factors like moisture, temperature, or mechanical fatigue, traditional composites frequently deteriorate at the interfaces. By eliminating distinct interfaces, FGMs lessen the possibility of delamination and crack spread, hence mitigating these problems [19]. Because of this feature, FGMs are especially well suited for challenging situations where durability is essential as in coastal or aircraft applications.

The ability to customize FGMs for certain uses is a noteworthy added feature. The versatility of traditional composites is limited by their reliance on set material combinations and stacking patterns. However, by regulating the gradations of features, FGMs can be customized to satisfy exact specifications.

To maximize thermal management, FGMs used in heat exchangers, for instance, can have low thermal conductivity at the heat sink end and high thermal conductivity at the heat source end [20]. In this context, FGMs can improve integration with biological tissues offering reliable performance in medical implants by combining biocompatible surfaces with strong and durable cores [21].

FGMs effectively mitigate residual stresses, which is a common problem seen in traditional composites. Warping, cracking, or early failure are often caused by residual stresses in composites that arise from mismatched behavior at interfaces during manufacturing or operational use. In order to overcome this issue, FGMs offer a gradual change in properties that lowers the accumulation of residual stresses which, in turn, improves structural stability [22].

Environmental and energy efficiency is one of the significant benefits of FGMs. Compared to traditional composites, FGMs are often produced using fewer adhesives and bonding agents, which diminishes production waste as well as its adverse impacts on the surroundings. Additionally, improved durability and thermal resistance result in energy savings in sectors where efficient thermal management is required, like power plants and automobile engines [23]. This is consistent with engineering's high demand on sustainable materials.

Recent developments in manufacturing methods have further enhanced the benefits of FGMs. Powder metallurgy, additive manufacturing, and 3D printing make it easier to produce complicated geometries which were difficult to accomplish by traditional composites and to precisely manage the gradation of materials. The potential applications of FGMs have increased owing to these developments, especially in high-precision fields of biomedical and aerospace engineering [24].

Overall, FGMs provide many advantages over traditional composites, like better thermal resistance, multifunctionality, increased structural integrity, customized property distribution, greater lifespan, and less residual stresses. These are primary components of contemporary material science because of their adaptability for specific applications and enhancement in manufacturing techniques. FGMs are delighted to play a crucial role in industries varying from aerospace to healthcare, rendering innovation and sustainability in engineering outcomes as long as industries seek high-performance materials.

8.2 Recent Advances in the Development and Application of Functionally Graded Materials

Functionally graded materials represent a class of refined composite materials and are distinguished by their unique design with a progressive change in composition and qualities across their thickness or other direction. FGMs can bridge the gap between materials with radically different characteristics like metal and ceramics on account of this gradation. Design, production, and utilization of FGMs sloped rapidly during the last few years, focusing on the growing needs of the aerospace, biomedical engineering, energy, and automotive sectors. These achievements indicate how FGMs are rapidly accepted worldwide as a better option than traditional composites for mitigating issues like stress concentration, heat resistance, etc.

(i) Advances in manufacturing techniques
The growing technologies have been reflected on the biggest advancement of FGMs. Recent technologies such as additive manufacturing (AM), which provides earlier inaccessible control over material gradation and complicated geometries, have been replaced with conventional techniques like centrifugal casting and powder metallurgy. AM techniques like selective laser melting (SLM), direct energy deposition (DED), and fused deposition modeling (FDM) have made possible the manufacturing of complex FGMs with accurate property distribution [8, 25, 26]. For instance, DED has been used to produce FGMs with extraordinary mechanical and thermal properties for application in aerospace, where parts must tolerate high mechanical stresses and temperature gradients.

Moreover, multi-material FGMs can now be produced through advancements in gradient control over property during the fabrication process. Researchers have tailored FGMs with specialized properties for specific applications, including high toughness in the core and wear resistance on the surface, by combining different material phases into one single structure [15, 16]. Computer-aided design tools have further enhanced the efficiency and accuracy of FGM fabrication by combining AM techniques, which have decreased production costs and increased their industrial applicability.

(ii) Integration with nanotechnology
A significant advancement in the recent manufacturing method of FGMs is the integration of nanotechnology. Researchers have remarkably improved

the mechanical strength, electrical characteristics, and thermal conductivity of FGMs by incorporating nanomaterials like graphene, carbon nanotubes, or nanoceramics [26, 27]. These nanocomposites opened up a new era of FGMs in fields like electronics cooling systems and thermos electric devices that demand special thermal management.

Additionally, the design of FGMs with hierarchical structures has been rendered easier by nanotechnology, allowing for a different property gradation both at the macroscopic and microscopic levels. High-precision applications in sensors, innovative energy systems, and microelectronics benefit greatly from such designs [8, 28]. For instance, by maximizing the gradation of thermal and electrical properties, nanostructured FGMs have been utilized in thermoelectric generators to increase energy conversion capacity [27].

(iii) Biomedical applications

Their capacity to replicate the natural gradations found in biological tissues has made FGM one of the most promising applications in biomedical engineering. Conventional implants frequently have problems like inadequate integration with neighboring tissues and stress shielding. By providing a progressive change in mechanical characteristics that closely mimic natural bone or cartilage, FGMs improve load distribution and lower the risk of implant rejection [21, 29].

FGMs with intricate geometries and biocompatible compositions customized for particular medicinal purposes can now be fabricated through recent developments in 3D bioprinting techniques [26, 28]. For instance, graded porosity titanium-based FGMs have been designed to promote bone ingrowth while retaining mechanical stability. In a similar vein, dental implants composed of FGMs that gradually transition from a biocompatible surface to a tough core have demonstrated better performance than those constructed of conventional materials [29].

(iv) Applications in energy systems

Developments in FGM have also made significant contributions to the energy sector, especially in the domains of heat exchangers, fuel cells, and thermoelectric devices. FGMs are ideal for applications demanding effective energy conversion and storage because of their capability to maximize thermal and electrical characteristics within a single component. Adjusting the gradation of thermal as well as electrical conductivities, current research suggest that FGMs can be used in thermoelectric generators to obtain greater energy conversion efficiencies [20, 27].

Additionally, FGMs have been used in solid oxide fuel cells (SOFCs) to enhance durability and performance in harsh operating environments. Researchers have boosted the thermal and mechanical stability of SOFCs by grading the characteristics of the electrode and electrolyte layers, resolving a major obstacle to their commercialization [15, 28].

(v) Environmental sustainability
Researchers are focusing on environmentally conscious materials and production strategies as sustainability has emerged as a key area of focus in the development of FGMs. For use in the construction and automotive industries, bio-based FGMs that integrate natural fibers and biodegradable polymers have been developed [26, 28]. By promoting recyclability and minimizing waste, these materials not only lessen the ecological impact of production but also adhere to the values of the circular economy.

By optimizing material utilization and eliminating the need for crucial physical prototyping, advanced simulation and modeling techniques have further contributed to the environmental sustainability of FGMs. By using these techniques, researchers can forecast how well FGMs will perform in different scenarios, ensuring that resources are used efficiently throughout manufacturing [24, 27].

8.3 Challenges and Future Directions

The difficulties in implementing functionally graded materials (FGMs) widely, especially those associated with manufacturing complexity and expense, are in line with the state of numerical research in this field at present. Although they have greatly aided in the study and optimization of FGM behavior, numerical approaches such as isogeometric analysis (IGA) or the finite element method (FEM) also draw attention to the limits imposed by manufacturing constraints.

The design of FGMs has been made easier by numerical models that forecast fracture mechanics, stress, and thermal performance. The accuracy needed to achieve graded material distributions, however, makes it difficult to translate these sophisticated models into workable manufacturing procedures, especially for large-scale production [8, 15]. For instance, techniques like IGA, which provide greater continuity and accuracy, have been widely used to model FGM structures; nevertheless, they necessitate manufacturing procedures capable of reproducing such accuracy [30].

While numerical tools such as FEM and IGA optimize material usage and reduce waste, the high costs of advanced manufacturing process like additive manufacturing (AM) limit their applicability [15]. Studies have revealed that numerical simulations, combined with cost–benefit analysis, can help in identifying the most economical gradation patterns for specific usages and perhaps reducing the barrier of entry for specific industries [8].

A revolutionary strategy for getting beyond these obstacles is the combination of machine learning (ML) with artificial intelligence (AI). The datasets produced by numerical techniques can be used by AI-driven algorithms to find the best designs and forecast production results. By quickly forecasting stress distributions for different gradation patterns, for instance, ML models can supplement FEM and drastically cut down on computational time and expenses [24, 27]. The design-to-production process can be streamlined with this method, increasing productivity and lowering resource requirements.

Due to its significant development in commercial packages and its modifications by industry and academics with substantial libraries, FEM may be a more effective and advantageous numerical technique than others [31]. In the following part, the finite element approach is used to examine the dynamic behavior of a functionally graded plate structure.

8.4 Free Vibration Analysis of Functional Composite Materials

8.4.1 Geometry and Property

Functional composite materials or functionally graded materials (FGM) are an advanced class of composites, the concept of which was proposed by a Japanese scientist in 1984. The tailored properties of two constituents (generally metal and ceramic) follow some mathematical model like power law, exponential law, sigmoid law, etc. The upper region is generally kept ceramic-rich to bear thermal stress, and the lower region is kept metal-rich for achieving high strength as shown in Figure 8.2.

The properties (Table 8.1) like Young's modulus of elasticity, density, Poisson's ratio, etc., may vary in between these two regions according to power law or sigmoid law distribution as shown in Figure 8.3. The mid-surface of the plate structure is aligned with the x-axis coordinate, and

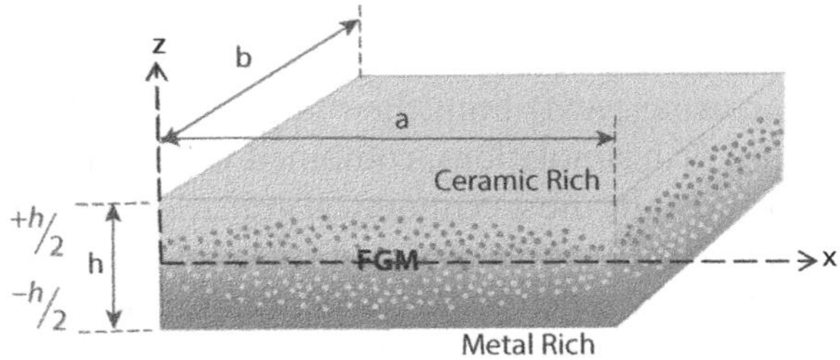

Figure 8.2 Schematic diagram showing the geometry of a rectangular FGM plate with ceramic-rich top and metal-rich bottom surface.

Table 8.1 Different mechanical properties considered for aluminum and zirconia.

Materials	**Young's modulus of elasticity, *E* (GPA)**	**Density, ρ ($^{kg}/_{m^3}$)**	**Poisson's ratio, ν**
Al (metal)	70	2,707	0.3
ZrO_2 (ceramic)	200	5,700	0.3

the volume fraction of the ceramic can be computed by using power and sigmoid law as given in Equations 8.1 and 8.2, respectively.

$$V_c^k = \left(0.5 + \frac{z}{h}\right)^n, V_m^k = 1.0 - V_c^k \tag{8.1}$$

In which, $-\frac{h}{2}(\text{metal}-\text{rich}) \leq z \leq +\frac{h}{2}(\text{ceramic}-\text{rich})$

For sigmoid distribution law, the volume fraction can be computed as

$$\begin{aligned} V_c^k &= 1.0 - \frac{1}{2}\left(\frac{h/2 - z}{h/2}\right)^n \text{ for } 0 \leq z \leq \frac{h}{2} \\ V_c^k &= \frac{1}{2}\left(\frac{h/2 + z}{h/2}\right)^n \text{ for } -\frac{h}{2} \leq z \leq 0 \end{aligned} \tag{8.2}$$

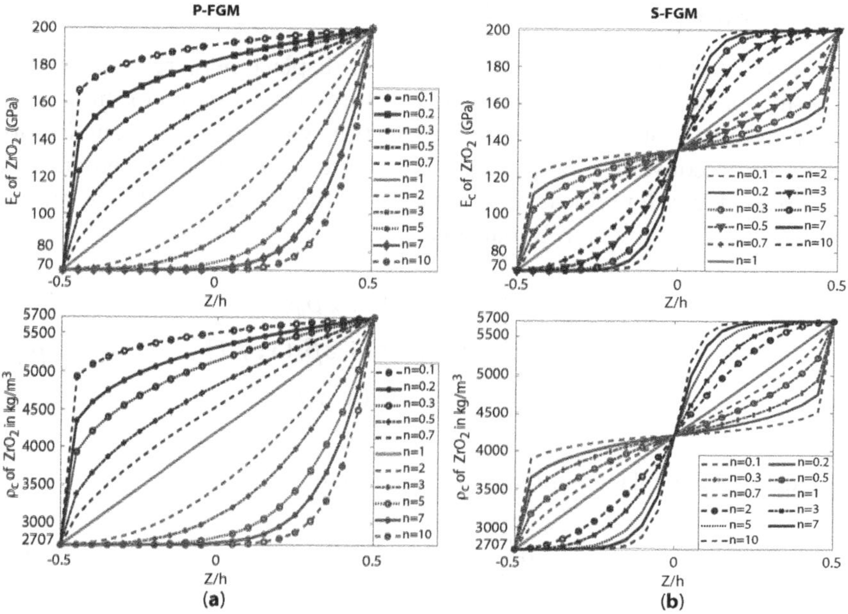

Figure 8.3 Distribution of elastic modulus and density of zirconia as a ceramic according to (a) power law and (b) sigmoid law.

Here V^k stands for volume fraction of any k-th layer along thickness direction and c and m stand for ceramic and metal, respectively. The power law index, often known as volume fraction index 'n' is a key parameter to determine a mechanical response which, when equal to zero, signifies complete ceramic and complete metal for infinite values.

The material's effective properties along the direction of thickness can now be computed by Voigt law as

$$P_{eff}(z) = (P_c - P_m)V_c(z) + P_m \tag{8.3}$$

where P stands for property. The value of $n = 1$ signifies that the region is equally captured by metal and ceramic, when $n < 1$, the region becomes more ceramic-rich and higher values of n (n >1) signifies a richer metallic region ($-0.5 \leq Z/h \leq 0$).

8.4.2 Mathematical Modeling

The displacement field and thereby strain–displacement relations which account for deformation of the plate is assumed to be based on first-order shear deformation theory (FSDT). The transverse displacement is supposed to be constant along the thickness and in plane displacement as a thickness-dependent linear function. Hence, the displacement field may be expressed as that given in Equation 8.4.

$$\begin{aligned} u_1(x,y,z,t) &= u(x,y,t) + z\beta_u(x,y,t) \\ u_2(x,y,z,t) &= v(x,y,t) + z\beta_v(x,y,t) \\ u_3(x,y,z,t) &= w(x,y,t) \end{aligned} \tag{8.4}$$

Hence, the displacement vector can be mentioned as

$$d^T = [u\ v\ w\ \beta_u\ \beta_v] \tag{8.5}$$

Here u, v, w are mid-plane displacements and β_u, β_v rotations about y and x axis respectively. The strain–displacement relations based on these presumptions are

$$\varepsilon = \begin{Bmatrix} \varepsilon_x \\ \varepsilon_y \\ \gamma_{xy} \end{Bmatrix} = \begin{Bmatrix} \varepsilon_x^{(0)} \\ \varepsilon_y^{(0)} \\ \gamma_{xy}^{(0)} \end{Bmatrix} + z \begin{Bmatrix} \varepsilon_x^{(1)} \\ \varepsilon_y^{(1)} \\ \gamma_{xy}^{(1)} \end{Bmatrix} = \begin{Bmatrix} \dfrac{\partial u}{\partial x} \\ \dfrac{\partial v}{\partial y} \\ \dfrac{\partial u}{\partial y} + \dfrac{\partial v}{\partial x} \end{Bmatrix} + z \begin{Bmatrix} \dfrac{\partial \beta_u}{\partial x} \\ \dfrac{\partial \beta_v}{\partial y} \\ \dfrac{\partial \beta_u}{\partial y} + \dfrac{\partial \beta_v}{\partial x} \end{Bmatrix} \tag{8.6}$$

$$\gamma = \begin{Bmatrix} \gamma_{yz} \\ \gamma_{xz} \end{Bmatrix} = \begin{Bmatrix} \gamma_{yz}^{(0)} \\ \gamma_{xz}^{(0)} \end{Bmatrix} = \begin{Bmatrix} \beta_v + \dfrac{\partial w}{\partial y} \\ \beta_u + \dfrac{\partial w}{\partial x} \end{Bmatrix} \tag{8.7}$$

The terms ε and γ refer to linear and shear strain, respectively, and the suffix (0) stands for mid-plane of the plate structure. The two-dimensional mid-plane is discretized into equal numbers of square elements along x and y axes as nine node quadrilateral elements. Five degrees of freedom $(u, v, w, \beta_u, \beta_v)$ are assigned to each node of an element. In each element, the geometric as well as displacement nodes are kept equal in numbers to retain the element isoparametric as shown in Figure 8.4. The shape function (N_i) for the nine-node Lagrange element can be articulated in terms of local coordinates with the following set of equations.

$$
\begin{aligned}
N_i &= \frac{1}{4}\xi\eta(\xi+\xi_i)(\eta+\eta_i), \qquad (i=1,2,3,4)\\
N_i &= \frac{1}{2}\eta(1+\xi)(1-\xi)(\eta+\eta_i), \qquad (i=5,7)\\
N_i &= \frac{1}{2}\xi(\xi+\xi_i)(1-\eta)(1+\eta), \qquad (i=6,8)\\
N_i &= \frac{1}{4}(1+\xi)(1-\xi)(1+\eta)(1-\eta), \quad (i=9)
\end{aligned} \tag{8.8}
$$

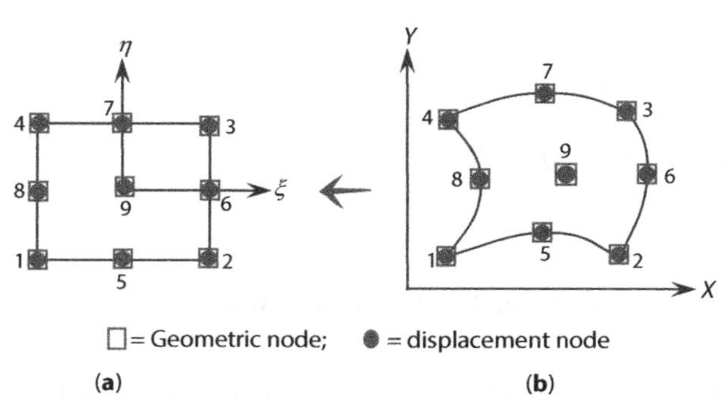

Figure 8.4 (a) Isoparametric nine-node quadrilateral element in local coordinate and (b) in a global coordinate system.

Hence, the displacement parameters in terms of shape functions can be formulated as:

$$u = \sum_{i=1}^{n} N_i(\xi,\eta)\hat{u}_i, \; v = \sum_{i=1}^{n} N_i(\xi,\eta)\hat{v}_i, \; w = \sum_{i=1}^{n} N_i(\xi,\eta)\hat{w}_i,$$
$$\beta_u = \sum_{i=1}^{n} N_i(\xi,\eta)\hat{\beta}_{ui}, \; \beta_v = \sum_{i=1}^{n} N_i(\xi,\eta)\hat{\beta}_{vi} \tag{8.9}$$

and

$$\begin{Bmatrix} N_{i,\xi} \\ N_{i,\eta} \end{Bmatrix} = \begin{bmatrix} J_{11} & J_{12} \\ J_{21} & J_{22} \end{bmatrix} \begin{Bmatrix} N_{i,x} \\ N_{i,y} \end{Bmatrix} = J_1 \begin{Bmatrix} N_{i,x} \\ N_{i,y} \end{Bmatrix} \Rightarrow \begin{Bmatrix} N_{i,x} \\ N_{i,y} \end{Bmatrix} = J_1^{-1} \begin{Bmatrix} N_{i,\xi} \\ N_{i,\eta} \end{Bmatrix}$$
$$= \begin{bmatrix} J_{11}^{*} & J_{12}^{*} \\ J_{21}^{*} & J_{22}^{*} \end{bmatrix} \begin{Bmatrix} N_{i,\xi} \\ N_{i,\eta} \end{Bmatrix} \tag{8.10}$$

in which J is the first Jacobian matrix for transforming natural to Cartesian coordinates and J^* is the inverse of it. The governing equation of the plate is formed by using Hamilton's principle in which the total strain energy and kinetic energy of the plate may be defined as follows:

$$U^e = \frac{1}{2} d^{eT} \int_{\Omega^e} \Big\{ \acute{B}_m^{(e)T} A \acute{B}_m^{(e)} + \tilde{B}_b^{(e)T} B \acute{B}_m^{(e)} + \acute{B}_b^{(e)T} B \tilde{B}_b^{(e)} + \tilde{B}_b^{(e)T} D \tilde{B}_b^{(e)}$$
$$+ \grave{B}_s^{(e)T} A_s \grave{B}_s^{(e)} \Big\} d\Omega^e d^e \tag{8.11}$$

which yields the stiffness matrix (K) as follows:

$$K^e = \int_{-1}^{1} \int_{-1}^{1} \Big\{ \acute{B}_m^{(e)T} A \acute{B}_m^{(e)} + \tilde{B}_b^{(e)T} B \acute{B}_m^{(e)} + \acute{B}_m^{(e)T} B \tilde{B}_b^{(e)} + \tilde{B}_b^{(e)T} D \tilde{B}_b^{(e)}$$
$$+ \grave{B}_s^{(e)T} A_s \grave{B}_s^{(e)} \Big\} |J| d\xi d\eta \tag{8.12}$$

in which

$$\acute{B}_m^{(e)} = \begin{Bmatrix} \frac{\partial N}{\partial x} & 0 & 0 & 0 & 0 \\ 0 & \frac{\partial N}{\partial y} & 0 & 0 & 0 \\ \frac{\partial N}{\partial y} & \frac{\partial N}{\partial x} & 0 & 0 & 0 \end{Bmatrix}, \tilde{B}_b^{(e)} = \begin{Bmatrix} 0 & 0 & 0 & \frac{\partial N}{\partial x} & 0 \\ 0 & 0 & 0 & 0 & \frac{\partial N}{\partial y} \\ 0 & 0 & 0 & \frac{\partial N}{\partial y} & \frac{\partial N}{\partial x} \end{Bmatrix},$$

$$\grave{B}_s^{(e)} = \begin{Bmatrix} 0 & 0 & \frac{\partial N}{\partial x} & 1 & 0 \\ 0 & 0 & \frac{\partial N}{\partial y} & 0 & 1 \end{Bmatrix} \tag{8.13}$$

$$\begin{aligned} \mathrm{A} &= \int_{-h/2}^{h/2} \mathrm{D_m}(z)\mathrm{d}z \quad & \mathrm{B} &= \int_{-h/2}^{h/2} z\mathrm{D_m}(z)\mathrm{d}z \\ \mathrm{D} &= \int_{-h/2}^{h/2} z^2\mathrm{D_m}(z)\mathrm{d}z \quad & \mathrm{A}_s &= \int_{-h/2}^{h/2} \mathrm{D_s}(z)\mathrm{d}z \end{aligned} \tag{8.14}$$

$$\begin{aligned} \mathrm{D_m}(z) &= \frac{E(z)}{1-\nu^2}\begin{bmatrix} 1 & \nu & 0 \\ \nu & 1 & 0 \\ 0 & 0 & (1-\nu)/2 \end{bmatrix} \\ \mathrm{D_s}(z) &= \frac{k_s E(z)}{2(1+\nu)}\begin{bmatrix} 1 & 0 \\ 0 & 1 \end{bmatrix} \end{aligned} \tag{8.15}$$

ν is Poisson's ratio and k_s is the shear correction factor $\left\{= 5(1+\nu)\big/(6+5\nu)\right\}$ [32] for FSDT. The kinetic energy ($\mathcal{K}$) of the FGM plate from which the mass matrix (M) is computed may be expressed as

$$\mathcal{K} = \frac{1}{2}\int_V \dot{d}^T \rho(z)\dot{d}dV \; ; M^e = \int_{-1}^{1}\int_{-1}^{1} N^T M N |J| d\xi d\eta \tag{8.16}$$

ρ is the density of the material. After incorporating, Hamilton's principle formulates the free vibration problem of the functionally graded material plate as

$$\{[K]^e - \omega^2[M]^e\}d^e = 0 \tag{8.17}$$

ω is the eigen value that determines the natural frequency of the plate.

The following boundary conditions are implemented on the functionally graded plate to study and compare the fundamental frequencies against different power law indices, aspect ratios, and slenderness ratios: (i) all the edges of the plate are simply supported (ssss), (ii) all the edges are clamped (cccc), (iii) two opposite edges are simply supported and another two are clamped (scsc), (iv) one edge is clamped, and others are free (cfff). Simply supported plates are applicable where edges are free to rotate without any bending moments. Clamped plates are suitable for roof structures, "scsc" plates may serve as a small bridge and "cfff" plates as cantilevers.

Table 8.2 First 10 non-dimensional frequencies of a square (b/a = 1, a/h = 20) ssss plate against different power law indices.

	$n = 0.5$		$n = 1$		$n = 5$		$n = 10$	
Mode no.	**IGA**	**Present**	**IGA**	**Present**	**IGA**	**Present**	**IGA**	**Present**
λ_1	5.593	5.4564	5.484	5.4451	5.593	5.3972	5.563	5.3315
λ_2	13.819	13.4733	13.549	13.4533	13.778	13.3407	13.701	13.1739
λ_3	13.819	13.4733	13.549	13.4533	13.778	13.3407	13.701	13.1739
λ_4	21.860	21.2988	21.432	21.2790	21.733	21.1094	21.609	20.8392
λ_5	27.125	26.4291	26.592	26.4140	26.919	26.2102	26.763	25.8695
λ_6	27.125	26.4291	26.592	26.4140	26.919	26.2102	26.763	25.8695
λ_7	34.883	33.9614	34.194	33.9589	34.529	33.7092	34.326	33.2618
λ_8	34.883	33.9614	34.194	33.9589	34.529	33.7092	34.326	33.2618
λ_9	37.971	36.3846	37.280	37.2800	35.176	37.0005	34.474	36.5670
λ_{10}	37.971	36.3846	37.280	37.2800	35.176	37.0005	34.474	36.5670

Table 8.3 First 10 non-dimensional frequencies of square (b/a = 1, a/h=20) cccc, scsc, and cfff plates at different power law indices.

Mode	cccc				scsc				cfff			
no.	$n = 0.5$	$n = 1$	$n = 5$	$n = 10$	$n = 0.5$	$n = 1$	$n = 5$	$n = 10$	$n = 0.5$	$n = 1$	$n = 5$	$n = 10$
λ_1	9.8025	9.7569	9.6768	9.5541	7.9305	7.8917	7.8254	7.7273	0.9677	0.9621	0.9535	0.9420
λ_2	19.6030	19.5266	19.3817	19.1261	14.8766	14.8088	14.6885	14.5017	2.3326	2.3204	2.3003	2.2719
λ_3	19.6030	19.5266	19.3817	19.1261	18.5857	18.5136	18.3711	18.1305	5.8628	5.8320	5.7819	5.7106
λ_4	28.3973	28.3084	28.1128	27.7305	25.1284	25.0400	24.8543	24.5239	7.4578	7.4193	7.3562	7.2652
λ_5	34.2273	34.1153	33.9103	33.4407	27.3315	27.2272	27.0208	26.6664	8.4055	8.3652	8.2958	8.1914
λ_6	34.4413	34.3519	34.1183	33.6472	33.6940	33.6043	33.3744	32.9148	12.3388	12.5987	12.7071	12.6456
λ_7	42.2745	42.1918	41.9241	41.3302	36.5110	36.6021	36.3475	35.8523	14.5171	14.4548	14.3405	14.1560
λ_8	42.2745	42.1918	41.9241	41.3302	36.7091	37.2800	37.0005	36.5670	16.5768	16.5041	16.3725	16.1629
λ_9	53.1832	53.1170	52.8066	52.0378	40.0195	39.9266	39.0307	38.8858	17.3465	17.2682	17.1301	16.9121
λ_{10}	53.1832	53.1170	52.8066	52.0378	44.5073	44.3839	43.3778	43.2283	19.0331	18.9543	18.8068	18.5634

The first 10 non-dimensional frequencies are validated with the IGA method [33] with a 16 × 16 mesh. The slenderness ratio (a/h) is kept equal to 20 and the aspect ratio (b/a) equal to 1 at simply supported boundary condition against different values of power law indices (Table 8.2). The non-dimensional frequency used is $\omega \frac{a^2}{h}\sqrt{\rho_c / E_c}$ [34] throughout the study. Table 8.3 presents a list of the first 10 non-dimensional frequencies at similar conditions for cccc, scsc, and cfff functionally graded plates [31].

As n increases from 1, the plate becomes metal-rich, with reduced stiffness and fundamental frequency. Conversely, as n decreases, the plate becomes ceramic-rich, increasing the stiffness and thereby increasing the fundamental frequency due to higher Young's modulus. Changing the boundary conditions from cfff to ssss, scsc, and cccc further increases the non-dimensional

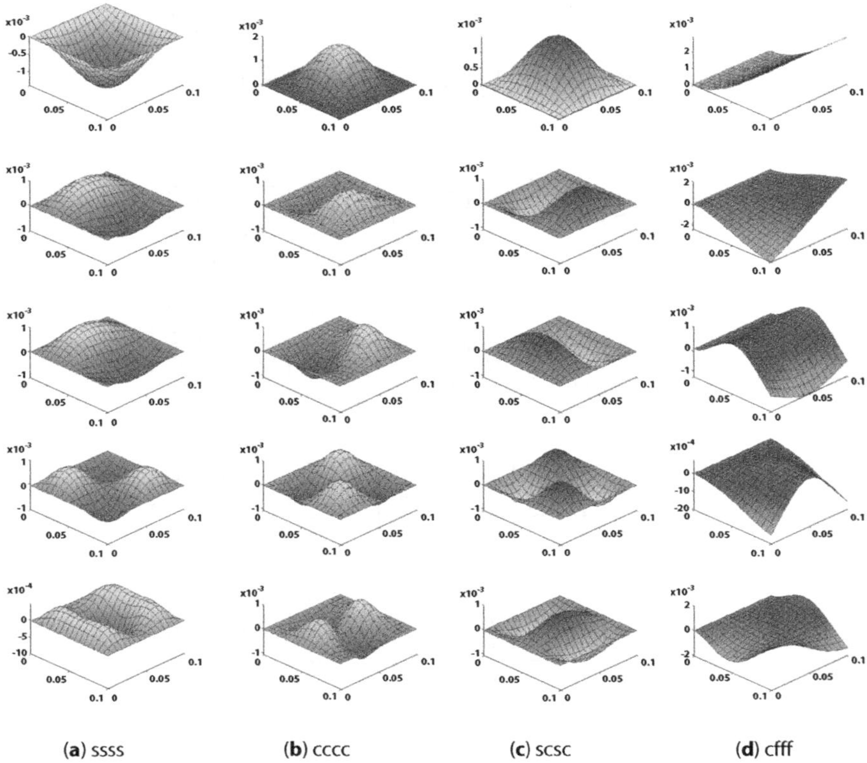

Figure 8.5 First five mode shapes for $n = 1$ of square FGM plate at boundary conditions (a) ssss, (b) cccc, (c) scsc, and (d) cfff.

frequency as structural stiffness improves. The mode shapes at all of the prescribed boundary conditions for the first five fundamental frequencies are shown in Figure 8.5. The effects of length-to-thickness ratios (5, 10, 20, and 50) on the first five non-dimensional frequencies at all of the mentioned boundary conditions and keeping n equal to 1 are shown in Figure 8.6. It is clear that with an increase in length-to-thickness ratio the frequency increases significantly up to a/h = 20 as the mass of the plate is reduced, keeping the stiffness high. Also, with high structural stiffness, the cccc plate yields the highest frequency and the cfff plate yields the lowest due to low structural stiffness. On the other hand, with an increase in aspect ratio, structural stiffness is reduced and thereby the frequency decreases except for the cfff plate in which due to the increase in lateral dimension along the clamped side, structural stiffness increases along that side, and this can be understood from Figure 8.7.

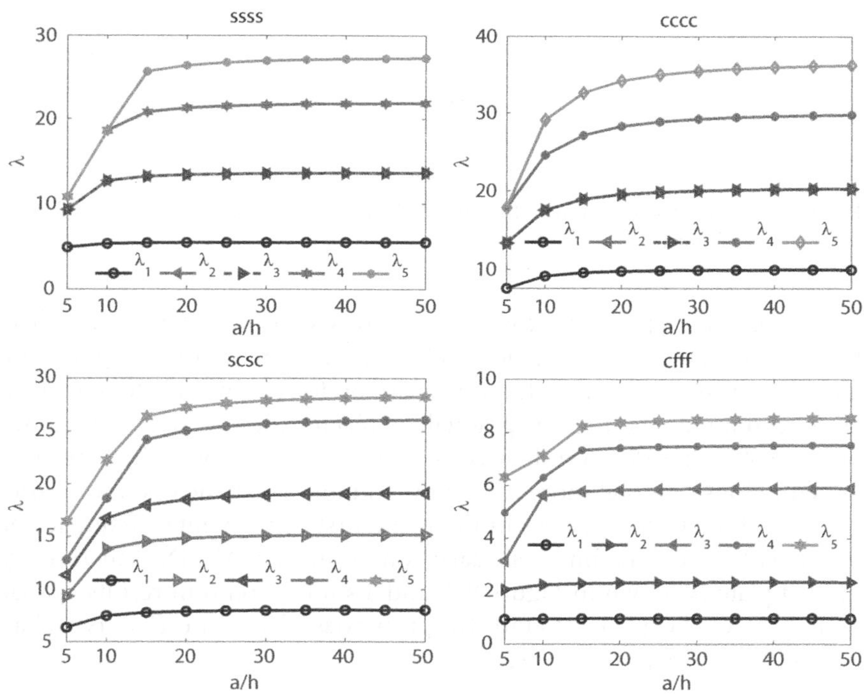

Figure 8.6 Effect of length-to-thickness ratio on the first five non-dimensional frequencies at different boundary conditions.

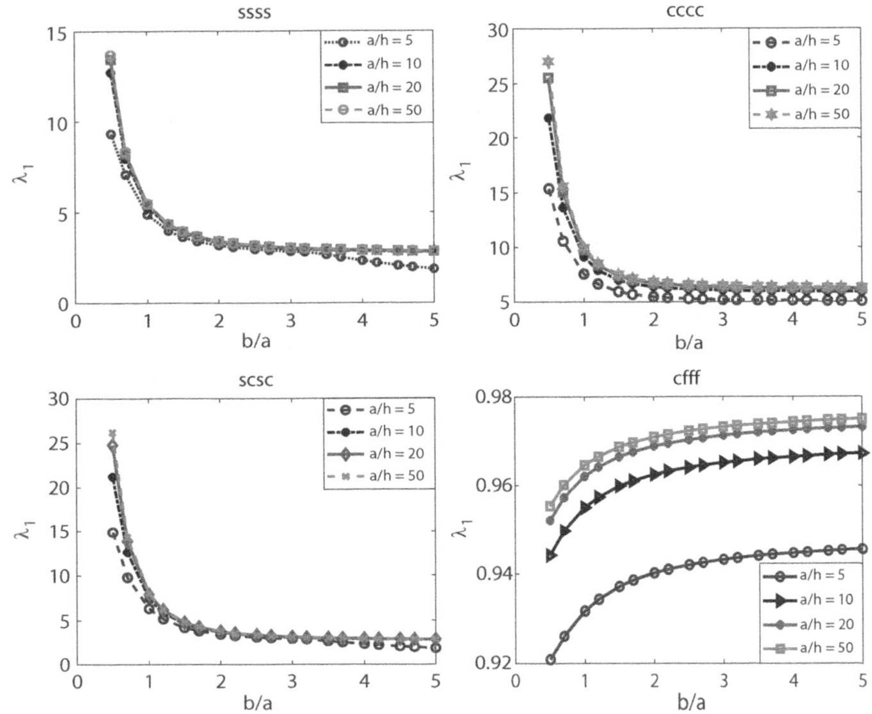

Figure 8.7 Effect of aspect ratio on the first fundamental frequency at $n = 1$ with different boundary conditions.

Owing to its superior mechanical and corrosion characteristics, such as their high modulus of elasticity, melting point, strength-to-weight ratio, and creep resistance, TiAl aluminides are preferred for high-temperature aircraft components. They are gradually replacing these heavier superalloys because of their 50% lower density than nickel-based alloys, which helps to cut emissions, use fuel more efficiently, and lower turbine operating costs [35, 36]. The effect of aspect ratio on the first fundamental frequency at $n = 1$ with all of the four boundary conditions for Al/TiN functionally graded plate is shown in Figure 8.8, and a similar trend in results as that with Al/ZrO_2 FG plate is found. The properties of TiN are $E_c = 490$ GPa, $\rho_c = 4{,}800$ kg/m^3, and $\nu = 0.3$.

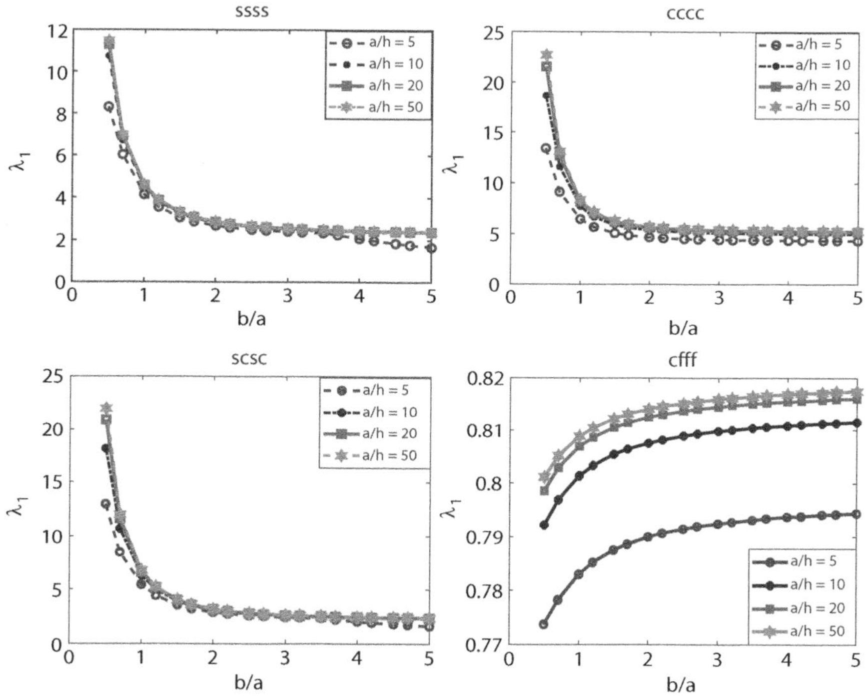

Figure 8.8 Effect of aspect ratio on the first fundamental frequency at $n = 1$ with different boundary conditions for the Al/TiN FG plate.

Similar to this, incorporating SiC ($E_c = 410$ GPa, $\rho_c = 3{,}100$ kg/m^3, $\nu = 0.22$) to aluminum-based FGM plates during fabrication [37, 38] is essential in enhancing mechanical properties. Because of its high strength and stiffness, low weight, and exceptional resistance to heat and wear, SiC is a great reinforcement material that may be used for a variety of structural components. For $n = 1$, b/a = 1, and a/h = 20, the first five natural frequencies of different FGM components with aluminum were examined (Figure 8.9). Al/SiC FGM plates exhibit the lowest fundamental frequency according to the results, which make them appropriate for low-frequency applications. Meanwhile, Al/Al_2O_3 and Al/TiN have nearly identical fundamental frequencies and TiN can replace Al_2O_3 where hardness and wear resistance are more crucial than toughness and shock resistance.

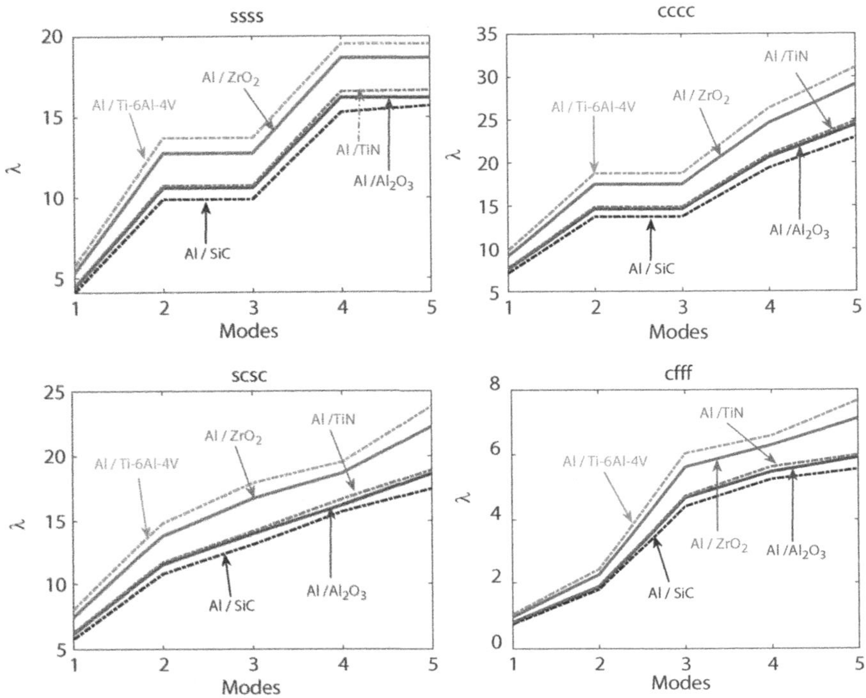

Figure 8.9 Comparison of fundamental frequencies at $n = 1$ for different boundary conditions of FGM materials.

8.5 Conclusion

While investigating free vibration in functionally graded material structures, the finite element method is a potential numerical tool that provides precise and effective solutions to the intricate mechanical behaviors of these systems. Under specified boundary conditions, FEM has been successfully used to investigate the parametric effects of power law index, length-to-thickness ratio, and aspect ratio on the fundamental frequency of FGM plates that are made of combinations of different ceramic materials with aluminum. Additionally, FEM supports development in FGM production techniques like additive manufacturing, centrifugal casting, and powder metallurgy. These procedures frequently provide graded characteristics and intricate geometries that are challenging to analyze with conventional techniques. With the help of FEM, these differences may be thoroughly studied, diminishing the chances of failure under vibrational stresses and establishing that designs fulfil performance requirements. As a result, FEM is a crucial numerical tool for maximizing the use of FGMs in engineering applications, bridging the gap between production techniques and realistic applications.

References

1. Barbero, E.J., *Advanced Composite Materials for Aerospace Engineering: Processing, Properties, and Applications*, pp. 1–413, Woodhead Publishing, Cambridge, United Kingdom, 2019.
2. Zweben, C. and Askeland, D.R., *Essentials of Materials Science and Engineering*, pp. 1–453, Cengage Learning, Boston, MA, USA, 2021.
3. Jamaluddin, M., Hasan, R., Alali, M.K., *Functionally Graded Materials in Design and Processing*, pp. 1–320, Elsevier, Amsterdam, Netherlands, 2021.
4. Kumar, A. and Kumar, R., Recent Developments in Functionally Graded Materials: Applications and Challenges. *Mater. Today: Proc.*, 45, 1021–1030, 2023.
5. Hashin, Z., Analysis of fiber composites. *J. Compos. Mater.*, 56, 4, 453–480, 2022.
6. Hanafi, I., Talib, R.J., Ahmed, M.M., Functionally graded materials in thermal barrier coatings: A review of applications and manufacturing methods. *J. Therm. Eng.*, 6, 3, 195–210, 2020.
7. Kim, J., Park, S., Lee, J., Mechanical and thermal properties of graphene-enhanced polymer nanocomposites: A review. *Compos. Part B: Eng.*, 190, 107948, 2020.
8. Zhang, M.Q. and Rong, M.Z., *Self-Healing Polymers and Polymer Composites*, pp. 1–430, John Wiley & Sons, Hoboken, NJ, USA, 2011.
9. Pimenta, S. and Pinho, S.T., Recycling carbon fibre reinforced polymers for structural applications: Technology review and market outlook. *Waste Manage. (Oxford)*, 105, 26–42, 2019.
10. Mishnaevsky, L., Brøndsted, P., Toftegaard, H., Recycling of composites: Strategies, technologies, and future challenges. *Compos. Struct.*, 300, 116115, 2022.
11. Jawaid, M. and Khalil, H.P.S.A., *Sustainable Composites for Aerospace Applications*, pp. 1–320, Woodhead Publishing, Cambridge, United Kingdom, 2021.
12. White, S.R., Sottos, N.R., Geubelle, P.H., *et al.*, Autonomic healing of polymer composites. *Nat. Mater.*, 19, 2, 235–245, 2020.
13. Sobczak, J.J., Drenchev, L., Drencheva, Z., Functionally graded materials: Opportunities and challenges in materials design and manufacturing. *Arch. Civ. Mech. Eng.*, 21, 2, 1–13, 2021.

14. Kieback, B., Neubrand, A., Riedel, H., Processing techniques for functionally graded materials. *Mater. Sci. Eng. A*, 362, 1–2, 81–105, 2003.
15. Matsunaga, H., Hirokawa, K., Nakamura, T., Thermal stress analysis of functionally graded materials subjected to heat flux. *Compos. Struct.*, 144, 180–188, 2016.
16. Shin, J., Kim, H., Lee, J., Mechanical properties of functionally graded materials with varying gradients. *Mater. Res. Express*, 5, 3, 035501, 2018.
17. Lahiri, D., Singh, V., Agarwal, A., Carbon nanotube reinforced aluminium composites: An overview. *Scr. Mater.*, 63, 11, 925–930, 2012.
18. Koizumi, M., The concept of functionally graded materials. *Ceram. Trans.*, 34, 3–10, 1993.
19. Birman, V. and Byrd, L.W., Modeling and analysis of functionally graded materials and structures. *Appl. Mech. Rev.*, 60, 5, 195–216, 2007.
20. Shi, X., Chen, L., Lin, Z., Thermal behavior of functionally graded materials in heat exchangers. *Therm. Sci. Eng. Prog.*, 10, 74–82, 2019.
21. Miyamoto, Y., Kaysser, W.A., Rabin, B.H., *Functionally Graded Materials: Design, Processing and Applications*, pp. 1–320, Springer, Berlin, Germany, 2013.
22. Hassan, M., Abid, M., Tariq, U., Residual stress reduction in functionally graded materials: A review. *J. Mater. Res. Technol.*, 9, 3, 4948–4963, 2020.
23. Khor, K.A. and Gu, Y.W., Effects of residual stresses on the performance of plasma sprayed functionally graded coatings. *Surf. Coat. Technol.*, 127, 1, 56–65, 2000.
24. Li, J., Song, X., Yu, X., Additive manufacturing of functionally graded materials: A review. *Procedia Manuf.*, 26, 774–785, 2018.
25. Li, J., Liu, Y., Wang, W., Machine learning for designing functionally graded materials: A comprehensive review. *Comput. Mater. Sci.*, 192, 110310, 2021.
26. Reddy, K., Kumar, V., Patel, R., Nanotechnology-enhanced functionally graded materials: Applications and advancements. *J. Mater. Res. Technol.*, 13, 2223–2235, 2021.
27. Singh, R.P., Kumar, A., Sharma, P., AI-driven innovation in functionally graded materials: From design to manufacturing. *Adv. Eng. Mater.*, 24, 6, 2101683, 2022.
28. Ahmed, A., Khan, M.I., Shah, S., Eco-friendly functionally graded materials for sustainable engineering applications. *Mater. Today: Proc.*, 47, 3145–3153, 2021.
29. Liu, W., He, Z., Xu, L., Functionally graded materials for biomedical applications: Advances and future perspectives. *Mater. Sci. Eng.: C*, 110, 110736, 2020.
30. Hughes, T.J.R., Cottrell, J.A., Bazilevs, Y., Isogeometric analysis: CAD, finite elements, NURBS, exact geometry, and mesh refinement. *Comput. Methods Appl. Mech. Eng.*, 194, 39–41, 4135–4195, 2005.

31. Gogoi, D., Pandit, M.K., Pradhan, A.K., Free vibration behavior in functionally graded material plates with dual efficient quadrilateral finite elements. *Mech. Based Des. Struct. Mach.*, *52*, 12, 9830–9857, 2024.
32. Timoshenko, S.P., On the transverse vibrations of bars of uniform cross-section. *Philos. Mag.*, 43, 125–131, 1922.
33. Van Do, V. and Lee, C., Free vibration analysis of FGM plates with complex cutouts by using quasi-3D isogeometric approach. *Int. J. Mech. Sci.*, 159, 213–33, 2019.
34. Hosseini-Hashemi, S., Fadaee, M. O. H. A. M. M. A. D., Atashipour, S.R., A new exact analytical approach for free vibration of Reissner–Mindlin functionally graded rectangular plates. *Int. J. Mech. Sci.*, *53*, 1, 11–22, 2011.
35. Gao, Y., Xie, Q., Gao, T., *et al.*, Design of functionally graded Ti–Al alloy with adjustable mechanical properties: a molecular dynamics insights. *J. Mater. Res. Technol.*, 23, 258–67, 2023.
36. Oh, S., Suh, D., Lee, S., Microstructure of TiN carbon steel functionally gradient materials processed by high-energy electron beam irradiation. *Mater. Sci. Eng.*, *A*, 248, 245–55, 1998.
37. Milosan, I., Bedő, T., Gabor, C., *et al.*, Characterization of aluminum alloy-silicon carbide functionally graded materials developed by centrifugal casting process. *Appl. Sci.*, 11, 1–14, 2021.
38. Vijaya Kumar, P., Jebakani, D., Velmurugan, C., *et al.*, Effect of SiC on Mechanical and Microstructural Characteristics of Al Based Functionally Graded Material. *Silicon*, 14, 1247–52, 2022.
39. Li, H., Wang, Z., Lv, H., Zhou, Z., Han, Q., Liu, J., Qin, Z., Nonlinear vibration analysis of fiber reinforced composite cylindrical shells with partial constrained layer damping treatment. *Thin-Walled Struct.*, 157, 107000, 2020.

9

Structural Health Monitoring of Composite Structures: Utility, Challenges, Sensor Technologies, and Advanced Methods

Sumanta Banerjee[1]* and Anindita Kundu[2]

[1]*Mechanical Engineering Department, Heritage Institute of Technology, Kolkata, West Bengal, India*
[2]*School of Computer Science and Engineering (SCOPE), Vellore Institute of Technology (VIT), Vellore, Tamil Nadu, India*

Abstract

Composite materials have revolutionized engineering applications due to their design flexibility, multi-functionality, superior strength-to-weight ratio, and corrosion resistance. However, their structural integrity can degrade over time due to structural damages like matrix cracking, fiber breakage, and delamination. This necessitates structural health monitoring (SHM) to ensure structural safety and integrity, enhanced service life, and optimum maintenance. This chapter begins by introducing traditional and smart composites and highlights their distinct characteristics and potential applications. It then discusses the challenges associated with composite materials, emphasizing the need for effective monitoring strategies. A comprehensive review of SHM techniques for composites is then presented, outlining their benefits and implementation challenges. Various sensor technologies, such as fiber optic sensors (FOSs), resistance strain gauges, eddy current sensors, piezoelectric sensors, and MEMS sensors, are explored for their applicability in SHM. Furthermore, an overview of key SHM methodologies is provided, categorized into "active" and "passive" approaches. Methods such as the Lamb wave technique, electro-mechanical impedance, active vibration-based monitoring, strain-based methods, acoustic emission, and comparative vacuum monitoring (CVM) are discussed. By offering insights into state-of-the-art SHM techniques and sensor technologies, this chapter serves as a knowledge resource

**Corresponding author*: sumanta.banerjee@heritageit.edu

Sandip Kunar, Pranav Charkha, Santosh Jaju and Harish Tiwari (eds.) Functional Composites: Role in Modern Engineering, (215–246) © 2025 Scrivener Publishing LLC

for researchers and engineers interested in the enhancement of reliability and longevity of composite structures.

***Keywords*:** Composite materials, structural health monitoring, SHM sensors, SHM methodologies

9.1 Introduction

9.1.1 Traditional Composites: Overview and Advantages

Recent advancements in manufacturing of multifunctional materials have opened up newer, unforeseen opportunities to create engineered structures and components that combine superior mechanical properties (e.g., tensile/shear strength, elasticity, toughness, fatigue strength, creep resistance, etc.) with a number of additional, companion capabilities (e.g., sensing, self-actuating, thermal and/ or electromagnetic functionalities). The objective is to fabricate structural components that integrate multiple functions, which enables the development of lighter and energy-efficient devices. Composite materials are particularly well suited for this purpose, owing to their multiphase composition and inherent customizability [1, 2].

Achieving multi-functionality in composite structures requires deep scientific foundation and an expanded understanding of the complex interactions among various physical phenomena at different length/time scales, which all contribute to these (novel) capabilities. Equipped with the desired (multi-functionality) attributes arising from the interplay of mechanical, thermal, and electromagnetic fields, composite materials offer significant possibilities for developing functionally adaptive, flexible structures [3, 4].

Composites are either naturally occurring or engineered materials and composed of two (or more) constituents that remain separate and identifiable, within the final structure, and with distinctly different physical or chemical properties. As a structurally integrated combination (of distinct materials) at the macroscopic level, (synthetic) composites are engineered for unique properties that are unattainable by their individual components in isolation. Unlike metallic alloys, each material in a composite retains its unique physical, chemical, and mechanical properties [1, 2, 5]. Composite materials consist of two primary phases: the *reinforcing phase* (in the form of particles, flakes, or fibers), which typically provides higher strength and stiffness, and the *matrix phase*, composed of ceramics, polymers, or metals.

These materials are commonly classified as fibrous composites, laminated composites, particulate composites, and mixed composites [6]. From the application perspective, composites can be categorized as "functional" or "structural" composites. While the former category exhibits novel physical, chemical, or electrochemical properties, the latter possesses distinctive mechanical properties. These materials are extensively used across a spectrum of applications, which include modern-day aircraft and spacecraft components, hulls of waterways transportation systems, sports gears and accessories, sensors and actuators, catalysts and metamaterials for active pollution control, biomedical materials, and batteries for "green energy" applications [7].

When compared with traditional metallic materials, composites offer a number of salient advantages enumerated as follows: (a) low density, which contributes to weight reduction, (b) higher values of specific strength (ratio of tensile strength and density) and stiffness (ability to resist deformation when subjected to an external force), (c) good vibration damping ability, (d) extended fatigue life (refers to the number of repeated and/or fluctuating stress cycles that a material can endure before fatigue failure), (e) high resistance to temperature, wear, corrosion, and creep (slow, transient deformation at high temperatures when subjected to constant load or stress), (f) customizability in microstructures and properties, which allows composites to be tailored for diverse applications, and (g) ability to integrate components into a single-cured assembly, which significantly reduce the need for fasteners and assembly labor [8].

The above-mentioned advantages have led to the widespread use of composite materials across various fields. In aeronautics (e.g., fighter jets, airplanes for commercial aviation) and astronautics (e.g., spacecrafts) applications, composites are increasingly employed for weight reduction, enhancement of payload capacity, and achieving operational range. In the automotive industry, the use of composites is growing to meet enhanced performance and fuel efficiency requirements. Wind turbine blades are typically made from composites in order to enhance (electrical) energy harvesting efficiency. In maritime and infrastructure applications, composite materials have received wide acceptance on account of high corrosion resistance. For components (e.g., brake and engine parts) subjected to high temperatures during operation, ceramic or metal matrix composites are commonly used. In addition, sports and recreation gears and accessories are emerging as a significant consumer of composite materials [8–11].

9.1.2 Smart Composites: Characterization and Application Potential

A "smart composite" is a "value-added" traditional composite (material) that is integrated with sensors, actuators, or responsive materials. These attachments enable it to sense and respond to environmental changes or external "stimuli" (e.g., thermal stress, moisture, or electric fields). These materials combine the structural benefits of traditional composites (as listed in the preceding paragraph) with advanced functional capabilities. The key features that characterize these "smart materials" are sensing capability, self-actuation, energy efficiency, and integration of novel components like shape memory alloys (SMAs), piezoelectric fibers, and magnetostrictive materials. SMAs can "remember" and revert back to their pre-defined shapes when subjected to mechanical and/or thermal stresses, leveraging their attributes of superelasticity (SE) and shape memory effect (SME) [12, 13]. Piezoelectric fibers are specialized materials that can generate electrical voltage in response to mechanical stresses or deformations. Conversely, they can deform mechanically, owing to the flexibility and lightweight nature of fiber structures, when subjected to external electric fields. Magnetostrictive materials undergo changes in shape or dimensions when exposed to external magnetic fields. Conversely, they can generate magnetic fields when mechanically deformed. This enables a two-way conversion between magnetic and mechanical energy [14, 15].

Smart composites, as we understand them today, are gaining prominence in scientific and technological progress in materials science and engineering. Recently, state-of-the-art advancements in design and optimized functionalities of composite structures have contributed to the evolution of smart materials and structures. Sustainable utilization of smart materials (in practical applications) necessitates reaching beyond the traditional rules of mechanics and demands interdisciplinary R&D expertise to understand, model, and predict their unique structural, signal-processing, and stimuli-sensing capabilities. The next generation of smart materials shall comprise an intricate network of interacting components and novel micro-structural morphologies; in addition, they shall exhibit unique, tailored responses to combined external influences (stimuli). Seamless integration of mechanical, thermal, electromagnetic, and other responses is essential, both at the material level and at the structural scale. At present, R&D efforts are focused on the design of futuristic "smart" materials equipped with inherent capabilities of structural health monitoring and/or non-destructive sensing, in addition to their capacity to bear mechanical loadings in specified applications. The aim is to synergize (support and enhance) these emerging technologies so that the latter function in a symbiotic manner [10, 11, 16, 17].

9.1.3 Structural Health Monitoring: Why the Need?

Structural health monitoring (SHM) involves periodic, or continuous, assessment of the integrity and performance of engineering structures. With the ever-rising use of composite materials and structures in a multitude of engineering applications (e.g., automotive, aerospace, civil, mechanical, and energy), SHM has gained prominence as a cutting-edge domain. Composites, notwithstanding their desirable attributes as durable, customizable, and lightweight, are prone to damage modes such as cracking, delamination, and fiber–matrix debonding, among others. The SHM protocols (for composites) ensures their reliable and safe operation, reduces maintenance costs, and extends their service life [18–20].

9.1.4 What This Chapter is All About?

This book chapter reviews the varied aspects of structural health monitoring for composite materials. The chapter begins with an "Introduction" section and outlines the significance of composites in engineering applications with their multifunctional attributes. An overview of the distinguishing traits between "traditional" and "smart" composites is presented. A discussion of the structural characterization of composites follows next. The advantages of the use of composites, over traditional materials, in varied engineering and technological applications are presented. The flow of ideas shifts focus, with a review of the characteristics and functionality of "smart" composites. The subsequent sections then discuss the necessity of structural health monitoring (SHM) and sets the stage for the chapter's focus. Following this, the "Composite Materials: Problems and Challenges" section highlights the limitations and challenges associated with composite materials, justifying a rationale for the implementation of SHM. The next section, "Structural Health Monitoring for Composites: Benefits and Challenges", explores the advantages that SHM offers in ensuring structural integrity while addressing the practical difficulties and challenges posed in its deployment. To provide deeper insights, the chapter presents a detailed review of sensors used in SHM and covers the functionalities and specifics of fiber optic sensors, resistance strain gauges, eddy current sensors, piezoelectric sensors, and MEMS sensors. The subsequent section, "Overview of Typical SHM Methods for Composite Materials", classifies the different SHM techniques into "active" and "passive" approaches and elaborates on the key methodologies. In line with published studies, overview of the Lamb wave method, the electro-mechanical impedance method, the active vibration-based method, the acoustic emission techniques, the strain-based methods, and the comparative vacuum monitoring (CVM) method is presented. The chapter concludes by summarizing the insights gained and discusses future directions for advancing SHM implementation in composite structures.

9.2 Composite Materials: Problems and Challenges

In spite of the advantages and wide applicability being offered, composite materials present some technical challenges. As multiphase materials, the properties of composites exhibit pronounced anisotropy. Their performance is dependent on the manufacturing process, which leads to variability in their (functional) capabilities. In addition, the mechanisms of flaw initiation and propagation within the composite matrix, as well as the modes of ultimate failure, are complex. Formulation of accurate models that describe the evolution of damage and fracture behavior in composites remains a challenging task. The underlying complexity of constitutive behavior and failure characteristics renders optimization of composite materials design a difficult task. Limited historical data and absence of established standards render the efficacy of composite designs heavily reliant on the designer's experience, which can result in over-designed structures [21]. Another notable area of concern is the susceptibility (of composites) to damages due to impact loading in the out-of-plane direction due to lack of reinforcements. Composites typically exhibit minimal total penetration under high-energy impacts. However, under medium or low (energy) impacts, cracking of the composite matrix can occur. These isolated cracks, under loading, can interact and trigger a delamination process. This phenomenon can induce fiber breakage preferentially on the side located opposite to the point of impact [22]. In addition, damages can result from aging (during functioning tenure) or faulty operations during manufacturing, assembly, or service [10, 21–23].

9.3 Structural Health Monitoring for Composites: Benefits and Challenges

Common types of damage in composites occur below the surface and are not readily visible and include (a) fiber breakage, (b) cracking of matrix, (c) debonding of fiber and matrix phases, and (d) delamination between constituent plies. These damages can potentially reduce the performance of composites, and their prompt detection can prevent sudden and total structural failures.

Traditional, non-destructive testing (NDT) procedures, such as ultrasonic testing, X-rays imaging, thermography, and eddy current-based fault detection are used to locate sites of cracks or fractures in composites. However, these methods are limited to offline, localized inspections and require complex, heavy equipment. For large-scale structures, these methods are found to be time-consuming and labor-intensive. In addition, disassembly of inspected structures, which may be necessary to access certain areas, elevates maintenance costs [24, 25].

SHM is an advanced technique evolved from non-destructive testing methods and integrates intelligent algorithms with state-of-the-art sensor technology to assess structural "health" condition of engineering structures [26]. However, unlike NDT, SHM leverages the potential of (locally placed) sensors for real-time and online damage detections. The need for SHM in composites has been steadily increasing due to its potential benefits, which includes enhanced reliability and safety, reduced lifecycle costs, and valuable insights for more reliable designs [27, 28].

Structural monitoring of composites presents several technical challenges, owing to (a) their complex behavior under applied stresses and (b) their heterogeneous structure. Unlike metals, composites do not exhibit uniform damage progression; instead, flaws often initiate at the microscopic level and propagate in unpredictable patterns. This complexity makes accurate model developments (for damage detection and evolution) prohibitively difficult. In addition, composites are highly sensitive to environmental factors (e.g., temperature, humidity, etc.), which hinder sensor performance and data interpretation. Lack of standardization of SHM protocols and variability in manufacturing processes of composites further impede the development of reliable monitoring systems. These challenges warrant advanced sensing techniques and robust methods of data analysis to ensure reliable SHM for composites.

The following sections introduce the sensors that are commonly used in SHM and review the common SHM methods for composites.

9.4 Sensors Employed in SHM: A Review

In SHM routines, sensors (of various categories) are embedded in inspected structures for collection of structure-related data (e.g., temperature, load-induced strains, and vibration). The commonly used SHM sensors include the fiber optic sensors, the resistance strain gauges, the eddy current sensors, the piezoelectric sensors, and the MEMS (micro-electromechanical systems) sensors. The following subsections provide brief descriptions of these different categories of sensors.

9.4.1 Fiber Optic Sensors (FOSs)

Fiber optic sensors make use of optical fibers to measure different parameters (e.g., strain, temperature, and vibration). Optical fibers are lightweight, and they are capable of providing precise measurements over long distances for their "immunity" against electromagnetic interference. Their ability to be embedded within structures suits SHM applications in composites. FOSs are suitable for SHM applications as they offer advantages of lightweight design, stability, reliable operations, longer lifespan, lower power consumption, immunity to electromagnetic interference (EMI), higher bandwidth, and compatibility with optical data processing and transmission. Based on their sensing range, the fiber optic sensors are classified as local, quasi-distributed, and distributed sensors [29]. The commonly used sensors in this category are interferometric sensors (e.g., the Mach–Zehnder interferometer, the Michelson interferometer, the Fabry–Perot interferometer, and the Sagnac interferometer). These sensors measure physical parameters by utilizing the principle of optical interference. These sensors detect changes in the intensity, wavelength, or phase of light as it propagates through, or interacts with, the sensors. These sensors either work on the (a) interference principle or (b) principle of phase shift detection. The former relies on the interference of two (or more) light waves. When light waves overlap, their phase difference determines whether the interference is constructive (whereby the intensity increases) or destructive (leading to decrease in intensity). The latter relies on the variations in the optical path lengths of one (or more) beam as a consequence of strains or displacements (external "stimuli"), which results in a measurable phase shift in interference patterns. The Mach–Zehnder interferometer splits the incoming light into two fragments, exposing their travel paths to different external conditions. The recombined light reveals changes in the phase due to (optical) path length differences. The Michelson interferometer is similar to Mach–Zehnder; however, it uses mirrors to reflect the light back and forms an interference pattern that is sensitive to changes in the optical path length. The Fabry–Perot interferometer utilizes multiple reflections between two closely spaced mirrors placed inside an optical cavity. Changes in the cavity length modify the interference of the component light waves, allowing for precise measurements. The Sagnac interferometer detects small-scale deformations and fractures on the basis of the Sagnac effect, whereby light waves that travel in opposite directions around a loop experience phase shifts due to rotation.

The Fiber Bragg grating (FBG) sensors are known for their multiplexing functionality, and represent a typical category of quasi-distributed fiber optic sensors. FBGs are created by introducing, along the core of a single-mode optical fiber, periodic modulations in the refractive index [30]. Constructive interference is observed when fiber path-constrained light passes through an FBG, occurring between the forward and the contra-propagating light waves, and produces a narrow-band back-reflection at the Bragg wavelength (λ_B). Any local changes in the FBG alter the value of λ_B, which permits the monitoring of strain-related parameters (resulting from mechanical stress or temperature) through the transmitted (or reflected spectrum) [31]. A notable advantage of an FBG sensor is its ability to deploy, within a single fiber, an array of wavelength-multiplexed FBGs for distributed measurements. Both time-division multiplexing (TDM) and spatial-division multiplexing (SDM) techniques can be implemented when the number of FBGs is increased. SDM involves arranging multiple FBGs at different positions along the same optical fiber, with each FBG characterized by a unique value of λ_B. When broadband light is passed through the (optic) fiber, each FBG reflects a specific λ_B, allowing the rest of the light to pass through to subsequent gratings. As the spatial separation ensures that each FBG reflects light at a distinct Bragg wavelength, it allows simultaneous measurements from all sensors along the fiber. The TDM involves assigning specific time slots for data acquisition from multiple FBG sensors. Delays are introduced between signals, from different sensors, along the same fiber. This is achieved by using a pulsed light source; as the light propagates along a fiber, reflections from each FBG are received at different times, owing to their varying distances from the source. This time delay is used to differentiate the signals from each FBG. Both SDM and TDM techniques enhance the capacity of FBG sensors, allowing multiple sensors to be integrated into a single optical fiber and improving efficiency in SHM systems [32].

Distributed fiber optic sensors (FOSs) utilize the entire length of an optical fiber as the sensing element, thereby enabling fully distributed measurements. This is particularly beneficial for monitoring large structures. These sensors operate by modulating the light intensity within the fiber. Two primary methodologies, namely, the optical time-domain reflectometry (OTDR) and Brillouin scattering, are employed for distributed sensing.

The OTDR technique is used to analyze the characteristics of an optical fiber by measuring the back-scattered light along its length. It is widely applied for fault detection in SHM systems. Herein this method, a short pulse of light is injected into the fiber. The back-scattered signal (the Rayleigh or Fresnel scattering) is then recorded as a function of time. The time delay of the back-scattered light aids in determining the location of faults, bends, or other (structural) anomalies. In contrast, the Brillouin scattering is a nonlinear optical phenomenon that leverages Doppler shifts in light frequency for measurements. Herein an incident light wave interacts with acoustic phonons (or density waves) in a medium. This leads to a frequency shift in the scattered light. This shift depends both on temperature and strain, thereby making it useful for distributed sensing. In this technique, a laser beam is sent in through the fiber, and the back-scattered Brillouin-shifted light is analyzed to determine changes in strain and temperature along the fiber length. The Brillouin optical time-domain reflectometry (BOTDR) and the Brillouin optical time-domain analysis (BOTDA) are two common sensing techniques that are based on this principle [10, 11, 33, 34].

9.4.2 Resistance Strain Gauges

Resistance strain gauges are instruments that are equipped to measure strains by detecting changes in electrical resistance in thin, conductive materials bonded to the surface of a structure. They are simple and reliable and also widely used for measuring localized deformations in SHM monitoring. However, their sensitivity to environmental factors (such as temperature) delimits their effectiveness in adverse conditions [35].

The resistance strain gauge is a conventional strain sensor and primarily consists of a foil or thin wire resistance grid, connectors, and an encapsulation layer. Utilizing the strain-resistance transducer effect, the grid detects strains in composite structures through changes in its resistance. The latter is converted into a voltage signal *via* a Wheatstone bridge circuit. With its minimal thickness and high sensitivity, the resistance strain gauge can be easily affixed to structures [10, 36, 37].

9.4.3 Piezoelectric Sensors

Piezoelectric sensors make use of the piezoelectric effect, wherein mechanical and/or thermal stresses generate an electrical potential difference (charge flow) in certain materials. They are widely used for monitoring vibrations, acoustic emissions, and dynamic strains.

Piezoelectric sensors are widely used in structural health monitoring of composites, on account of their ability to convert mechanical strains into electrical signals. Certain materials (e.g., quartz, lead zirconate titanate (PZT) ceramics wafers, or polyvinylidene fluoride (PVDF) films) generate an electric potential when subjected to mechanical deformations. Phenomenologically, the behavior of "charge response to a stress field" is called the piezoelectric effect. Conversely, inverse piezoelectric effect is observed when an externally applied electric field sets up an induced strain field within these materials. Hence, piezoelectric sensors can be employed as both sensors and actuators [38].

PZT wafers are characterized by a high value of piezoelectric constant, and they exhibit excellent sensitivity as sensors, with strong actuation capabilities. However, their ceramic nature renders them inherently brittle. In contrast, PVDF films are lightweight, cost-effective, flexible, and offer superior internal damping [39]. Their weak inverse piezoelectric properties make them more suited for sensing applications (rather than actuation) [40]. To address the brittle properties of PZT wafers, composites with piezoelectric properties (that include piezoelectric paints and rubbers) have been synthesized as viable, alternative solutions.

To summarize, piezoelectric sensors are sensitive, compact, and capable of detecting dynamic strains, pressure, and vibrations (at low as well as high frequencies). They are commonly used for acoustic emission monitoring (AEM), Lamb wave-based structural inspections, and vibration-based SHM techniques. This makes them effective for real-time damage detections in composite structures [10, 11].

9.4.4 Eddy Current Sensors

Eddy current sensors are non-destructive evaluation (NDE) tools used for structural health monitoring (SHM) of composites. These sensors operate on the principle of electromagnetic induction, where a time-varying magnetic field (generated by a coil) induces eddy currents in a conductive material. Composites, which often include conductive fibers like carbon, facilitate the formation of eddy currents, which then interact with the material's structure. Any changes due to structural damage (e.g., cracks, delamination, or fiber breakage) alter the eddy current flow. This change is detected by the sensor, thereby providing real-time information on structural integrity.

Eddy current sensors are advantageous for SHM due to their (a) high sensitivity to surface and near-surface defects, (b) ability to operate without direct contact, and (c) suitability for real-time monitoring. They are particularly effective for detecting damage in conductive composites, such as carbon fiber-reinforced polymers (CFRPs), and can be integrated into composite structures for continuous monitoring. However, their effectiveness is limited in non-conductive or poorly conductive composites, and they may struggle with deep subsurface defects. Overall, eddy current sensors are a valuable tool for ensuring the safety and longevity of composite structures in aerospace, automotive, and civil engineering applications.

The principle of electromagnetic induction forms the basis of using eddy currents for damage detection [41], where eddy currents are induced in the conductive matrix and detected by the same (or different) windings of the sensors. These sensors are primarily used for detecting cracks or corrosion in metallic components, even through non-conductive coatings or layers. This capability makes them particularly valuable for composite structures, such as metal-matrix composites and parent metal materials with composite doubler repairs.

Conventional eddy current sensors, owing to their bulky winding configurations, are difficult to integrate into structures. Presently, advanced micro-fabrication techniques can create windings that adhere to or can be directly printed on flexible substrates of composites. Eddy current foil sensors that include arrays of multiple sensing elements (of various shapes) have been developed [42]. These new sensors are thin, flexible, and can be easily surface-mounted or embedded between layers, which make them ideal for continuous, real-time, on-line and monitoring [10, 11].

9.4.5 Micro-Electromechanical (MEMS) Sensors

MEMS sensors have emerged as a promising technology for health monitoring of composite structures in aerospace, automotive, civil, and marine applications. Owing to their small size, low energy consumption rates, higher sensitivity, and cost-effectiveness, MEMS sensors are widely utilized for real-time monitoring of structural components. Enabling early detection of structural defects, these sensors ensure improved reliability, safety, and longevity of composite structures.

The MEMS sensors used in SHM include strain sensors, pressure sensors, acoustic sensors, accelerometers, and gyroscopes. MEMS strain sensors detect deformation in composite materials under mechanical loads, which enable early damage identification. MEMS acoustic sensors assist in capturing acoustic emissions generated by crack formation and fiber breakage in composites. MEMS accelerometers and gyroscopes are particularly useful in vibration-based SHM, wherein they measure changes in dynamic response due to structural defects.

MEMS sensors operate based on μ-fabricated structures that convert mechanical stimuli (acceleration, strain, or pressure) into electrical signals. These signals are processed, using advanced algorithms, for detection of structural anomalies. To enable *in-situ* real-time monitoring, these sensors can be embedded within, or surface-mounted on, composite structures. The benefits of wireless networking allows for continuous data collection and transmission to remote monitoring systems. These aspects reduce maintenance efforts and costs.

To summarize, MEMS sensors offer the following advantages:

- Miniaturization: Their small sizes allow for seamless integration into composite structures, without affecting mechanical properties appreciably.
- High sensitivity: MEMS sensors can detect even minute configurational changes, which makes them ideal for early damage detection.
- Low power consumption: The aspect of energy efficiency enables long-term monitoring with minimal power requirements.
- Wireless capability: MEMS-based SHM systems form wireless sensor networks, which eliminates the need for complex wiring and reduces installation costs.
- Cost-effectiveness: Mass production of MEMS devices lowers the overall cost of SHM implementation.

In spite of salient advantages, MEMS sensors face technical constraints such as (a) environmental sensitivity, (b) signal noise, and (c) power limitations. Environmental factors like temperature fluctuations and electromagnetic interference tend to affect sensor-accuracy adversely. Among others, improving the robustness and durability of MEMS sensors in harsh operating conditions remains a primal research focus.

Advanced integrated circuit (IC) fabrication techniques have enabled MEMS technology to integrate μ-sensors, actuators, and control functions into a single silicon chip. As an intelligent system, MEMS can detect environmental conditions, and respond accordingly through μ-circuit control in real time [43]. At present, various MEMS sensors, which include accelerometers and pressure sensors, are commercially available. Owing to their miniatured physical dimensions and high degree of integration, these sensors exhibit notable advantages such as (a) flexible configuration, (b) lightweight design, (c) low power consumption, (d) rapid response time, (e) minimal noise, (f) high reliability, and (g) cost efficiency. In addition, the need for long cables can be averted by embedding wireless communication capabilities; the latter can be achieved by embedding transmitter chips within the sensors.

In addition to comparative vacuum monitoring (CVM) sensors, laser scanners (a type of MEMS sensor) and microwave sensors can be employed for SHM of composites.

In order to cover larger monitoring areas with enhanced fault-detection capabilities, structural inspection leverages the advantages of extensive sensor arrays embedded within the monitored structures, which comprise of a large number of identical (or different types of) sensors arranged in pre-determined patterns [44]. This state-of-the-art development has led to the implementation of multiple layers of sensor arrays. These layers, with customized configurations, are fabricated by embedding sensor elements within thin, flexible dielectric films, and bear resemblance with eddy current arrays [45]. The advantages of these sensor layers include (a) enabling rapid and consistent deployment of a large number of sensors, (b) eliminating the need for connecting wires, thereby reducing the scale of electromagnetic interference (EMI), and (c) allowing the layers to be either embedded as additional layers in composite materials or surface-mounted on existing structures (during manufacturing).

Notable sensor array layers include the PZT array layer (commonly referred to as the SMART layer), the PZT-FOS hybrid array layer, and the HELP (hybrid electromagnetic performing) layer [46].

Integration of artificial intelligence (AI) and machine learning (ML) for data processing in MEMS technology will enhance their capabilities further. Using energy harvesting techniques, the development of self-powered MEMS sensors is expected to revolutionize SHM through autonomous monitoring of composite structures [10, 11].

9.5 Overview of Typical SHM Methods for Composite Materials

As schematically depicted in Figure 9.1, structural health monitoring can be categorized as either "active" or "passive". This classification depends on whether actuators are employed in the process, or otherwise [47]. In "passive" mode of SHM, operational parameters (e.g., stress, loads, acoustic emissions, and environmental conditions) are analyzed to assess real-time structural states. This assessment is achieved through signal processing techniques, system modeling and analysis, or *a priori* knowledge. "Passive" inspection methods "listen" to a structure without interacting with it; in contrast, actuators are used to excite the structure in a controlled manner in the "active" mode. The resulting structural responses are, thereafter, analyzed for assessment. It may be noted that "active" monitoring requires both actuators and sensors as and when needed.

In composites, common "active" SHM techniques are the Lamb wave method, the electro-mechanical impedance method, and the active vibration-based method. On the other hand, typical "passive" approaches include acoustic emission, strain-based monitoring, and the comparative vacuum monitoring method. Based on published studies, the following sections outline the chief attributes of each of these methods.

9.5.1 Lamb Wave Method (LWM)-Based SHM

Lamb waves (after H. Lamb; 1917), a type of guided ultrasonic waves, propagates in thin-walled structures. LWM is a prevalent SHM technique, for structural health monitoring of composites, as Lamb waves are (a) capable of long-distance propagation, and (b) highly sensitive to both surface and internal defects. This method leverages the phenomenon of

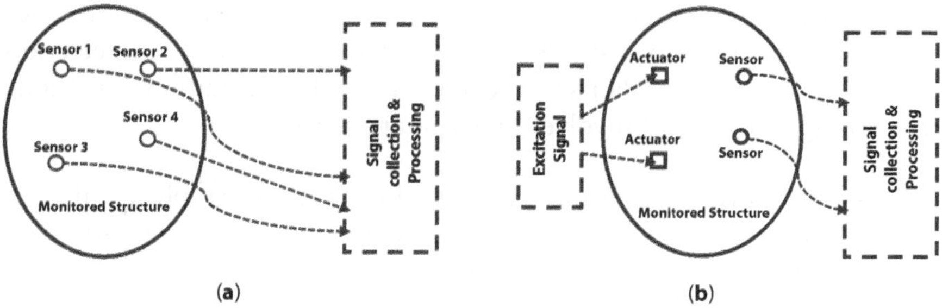

Figure 9.1 Schematic of (a) passive SHM and (b) active SHM.

elastic wave propagation for detection and characterization of structural damage. Lamb waves can propagate in thin plate-like structures, and consist of "symmetric" (S) and "asymmetric" (A) modes. Being sensitive to both material properties as well as structural defects, these waves are ideal for SHM of composite materials [48].

In this method, piezoelectric sensors or actuators (e.g., lead zirconium titanate or PZT wafers) are typically used to generate and detect Lamb waves. When the waves encounter a structural defect/damage (e.g., delamination, cracks, or debonding), they scatter or reflect. This causes changes in wave properties such as amplitude, phase, or velocity. These changes are then analyzed to locate and quantify the damage.

The propagation of multi-modal Lamb waves is rendered complex by their dispersion characteristics. From the wave propagation perspective, structural defects alter structural geometry and loading boundary conditions, which gives rise to reflection, scattering, and spatio-temporal attenuation of energy during waves-structure interactions. These interactions enable the extraction of characteristic parameters from the wave signals, which can be used for damage inspections.

The location of structural defects is typically determined by the "time-of-flight" (TOF) of wave signals. In the "ellipse location method", knowing the TOF of defect-induced scattered signals from a transducer pair and the wave traveling speed allows the creation of an ellipse contour as a probable damage locus, with the transducer pair at its foci. In order to pinpoint the exact location of defect, additional ellipses are constructed, using the signals scattered from other transducer pairs. These ellipses intersect; their intersection points pin down the flaw-site(s). Theoretically, at least three ellipses are required for precise location of damage. When significant mode conversion occurs during defect-induced scattering, the TOF of signals can be utilized to assess the fault location.

The time-reversal method (TRM) for damage detection is also an advanced technique in SHM that utilizes the principle of time-reversed wave propagation to detect and localize defects. This novel method relies on the wave reciprocity property, and is particularly effective in detection and characterization of damage in composites, where traditional methods may struggle. The TRM involves two primary steps:

- Wave propagation and recording: In this (forward) phase, an excitation signal, generated by an actuator (typically, a piezoelectric transducer), propagates through the structure. The waves travel through the structure, interact with the material boundaries, and scatter due to defects or discontinuities.

These scattered waves are, then, captured by sensors placed on the structure.
- Time-reversed playback: In this (reversed) phase, the recorded waveforms are time-reversed and re-transmitted into the structure by the same sensors that initially received the signals. The time-reversed waves retrace their original propagation paths back to the source, and converges at the defect location (if any such exists).

By analyzing the focal point and the signal intensities during the time-reversed playback, the presence and location of damage can be accurately identified.

To summarize, the salient features of the TRM are:

- High sensitivity: TRMs are highly sensitive to structural defects, which renders them ideal for detecting subtle damages in composites.
- Self-focusing: The time-reversed waves naturally focus back on the sites of defects, which offers precise spatial resolution for damage localizations.
- Minimal baseline dependency: Time-reversal techniques do not require extensive baseline data, which makes them more effective in operational environments.

The principal advantages of the Lamb wave method include its (a) ability to monitor large areas, (b) sensitivity to subsurface defects, and (c) compatibility with real-time SHM systems. However, the dispersive characteristics and (complex) multimodal nature of the Lamb waves warrant advanced signal processing techniques for accurate interpretation of fault-related information.

The delay-and-sum (DAS) imaging method is also a prevalent signal-processing technique in Lamb wave-based SHM. This simple and robust method relies on the principle that Lamb waves, when scattered by (structural) defects, generate reflections that can be captured by an array of sensors. The DAS algorithm processes these signals by applying time delays, which correspond to possible damage locations, and sums up the amplitude contributions coherently. This enhances the signal strength at the defect site and suppresses (signal) noise. The resulting damage index map (DIM) provides a visual representation of the defect location and severity. The effectiveness of the DAS method depends on factors such as (a) wave dispersion, (b) signal-to-noise ratio, and (c) distribution of sensors.

Phased arrays are also an advanced technique in this regard. A phased array system consists of multiple piezoelectric transducers (e.g., PZT) arranged in a specific geometric configuration. Exercising control over the phase and amplitude of the signals transmitted from each transducer, the Lamb waves can be steered and focused in desired directions. This enhances the detection of defects such as cracks, delaminations, and impact damage. The key advantage (of phased arrays) is the ability to electronically scan and focus wave energy by *stationary* sensors. This provides (a) better defect characterization, (b) higher resolution imaging, and (c) improved signal-to-noise ratio. By applying beamforming techniques, phased arrays facilitate directional wave propagation and selective mode excitation, which helps in enhancing damage-detection accuracy and minimize dispersion effects.

The tomography imaging method is yet another powerful technique used in Lamb wave-based SHM for locating faults and severities in structural composites. In line with medical imaging principles, this method employs multiple transducers (positioned at various locations) to send and receive Lamb waves. By analyzing changes in wave propagation characteristics, e.g., time of flight (TOF), amplitude attenuation, and phase shifts, a tomographic image of the internal structure is generated. This technique specializes in locating distributed defects, such as delaminations, matrix cracking, and impact-induced defects, by mapping variations in wave velocity and energy distribution. Suitable algorithms (e.g., inverse problem solving and iterative reconstruction techniques) are used to process the collected Lamb wave data and generate a high-resolution 'damage map'. As compared to (conventional) point-by-point fault-location methods, Lamb wave tomography offers (a) enhanced spatial resolution, and (b) ability to monitor larges areas with fewer sensors [10, 11, 49, 50].

9.5.2 The Electromechanical (E/M) Impedance Method

The E/M impedance technique is an effective tool for SHM of composites, which leverages the coupling between mechanical vibrations of composite structures and the electrical impedance of piezoelectric sensors. This method uses piezoelectric sensors (e.g., lead zirconate titanate or PZT), which serve as both actuators and sensors for monitoring changes in structural integrity.

The working principle is as follows: E/M impedance is based on the relationship between the electrical impedance of a piezoelectric sensor, and the mechanical impedance of the structure (to which the sensor is bonded). It may be noted that direct measurements for the mechanical impendence is difficult. Electrical impedance is a measure of resistance that a circuit (or a passive component) presents to an alternating current (AC). Analytically, it has a complex representation denoted by $Z_E = R + jX\ (\Omega)$, where "R" denotes "resistance" and "X" stands for "reactance". The "resistance" denotes the "real" part of impedance $\left(R \equiv Re(Z_E)\right)$, which opposes current flow irrespective of frequency. The "reactance" is the 'imaginary' part $\left(X \equiv Im(Z_E)\right)$, which arises due to capacitors and inductors in a circuit, and varies with frequency. As impedance is frequency-dependent, its magnitude changes with the frequency of AC signal. In piezoelectric SHM applications, Z_E is influenced by mechanical properties of the structure bonded to the sensor, which facilitates detection of damage by monitoring changes in the impedance. Mechanical impedance (Z_M) is a measure of resistance that a structure (or material) offers to motion, when subjected to an external force. It is defined as $Z_M = F/U$ (Ns/m), the ratio of the applied load (F) and resulting speed (U). Mechanical impedance includes contributions from the material characteristics (e.g., mass, stiffness, and damping). Similar to Z_E, Z_M is also frequency-dependent. In SHM, variations in Z_M (due to structural changes) alter the vibrational response of the structure. These changes can also be indirectly detected by monitoring the electrical impedance of piezoelectric sensors (influenced, in turn, by the structure's mechanical impedance). Applying an alternating voltage to the sensor induces vibrations in the structure. The structural response, influenced by its mechanical properties (e.g., stiffness, mass, and damping), alters the electrical impedance of the sensor. Any damage (such as cracks, delaminations, or voids) changes the local mechanical impedance, which can be detected through variations in the measured electrical impedance.

The advantages of the impedance method are (a) sensitivity to local damage, which makes it ideal for detecting defects at small length scales in composite structures, (b) monitoring at high, ultrasonic frequency ranges (typically above 30 kHz), which allows for detection of subtle changes in structural integrity, (c) real-time SHM, which enables immediate (dynamic) feedback on structural health, and (d) provisions for compact and lightweight piezoelectric sensors, which are easy to integrate with composite structures. Some limitations of the method are (a) local monitoring capabilities, which makes it effective only for local damage-detection, (b) requirement of multiple sensors for larger structures, which increase initial costs, (c) environmental sensitivity, which implies that variations in temperature and other environmental factors adversely affect the accuracy of impedance measurements (necessitating compensation techniques), and (d) calibration requirement, which requires baseline measurements of an intact structure for accurate assessment of damage.

In addition to detection of delaminations, voids, and cracks, the impedance method enables monitoring the strength of adhesive bonds and joints in composite assemblies. It also serves to assess the (structural) integrity of large panels or complex geometries [10, 51].

9.5.3 Active Vibration-Based Method (AVBM)

The AVBM is a classical monitoring technique, based on the principle that dynamic characteristics of a structure are directly influenced by its physical properties (e.g., mass, stiffness, and damping). It involves exciting a composite structure with controlled vibrations and analyzing the response (of structure) for detection and evaluation of damage. Any damage that alters these physical properties will result in measurable changes in the vibration response. This approach, therefore, leverages the fact that damage (e.g., matrix cracking, delamination, or fiber breakage) alters the dynamic response of the structure, which includes natural frequencies of vibration, frequency response function (FRF), power spectrum, mode shapes, mode curvature, mode flexibility matrix, energy transfer rate (ETR), and damping characteristics. The working principle is briefly as follows: External actuators (e.g., piezoelectric transducers or shakers) induce vibrations in the structure. Thereafter, sensors (e.g., accelerometers or piezoelectric sensors) measure the structural response to the vibrations. The resulting changes in dynamic properties, such as shifts in natural frequencies, distortions in mode shapes, or variations in damping, are analyzed to identify and locate damage.

An active vibration-based SHM system (for composites) typically consists of the following key components and processes:

- Excitation mechanism: An external actuator is used to introduce controlled vibrations into the composite structure. Common structure-integrated excitation mechanisms include (a) piezoelectric actuators, which convert electrical energy into mechanical vibrations, (b) electromagnetic shakers that provide precise, localized vibrations over a wide frequency range, and (c) impact hammers, used for generating transient vibrations through localized impacts.
- Response sensors: Sensors are installed on the structure to measure its response to the applied vibrations. The common types of response sensors include (a) accelerometers, for measuring vibration accelerations and ideal for global monitoring, (b) piezoelectric sensors, which detect strain

(or displacement) that are caused by vibrations, and (c) laser vibrometers, for providing non-contact measurement of surface vibrations.

- Data acquisition system (DAQS): The DAQS digitizes and processes the signals collected from the sensors. Principal components of a DAQS include (a) signal amplifiers to ensure accurate readings, (b) analog-to-digital converters for transforming sensor outputs into digital data, and (c) high-speed processors for real-time data handling.
- Signal processing and analysis: The collected data is processed to extract dynamic parameters of the (composite) structure, such as natural frequencies, damping ratios, modal shapes, frequency response functions (FRF), and energy transfer rates (ETR). Changes in these parameters are analyzed to detect damage. Advanced techniques like Fast Fourier Transform (FFT), wavelet analysis, or machine learning algorithms are employed in state-of-the-art systems to enhance sensitivity and accuracy.
- Damage detection and diagnosis: Structural damages are detected by (a) comparing the current vibration response to a baseline model that represents the undamaged structure and (b) theoretical predictions from structural models. Significant deviations indicate potential instances of damage, and localization techniques (e.g., mode shape curvature or energy transfer mapping) can pinpoint the affected area.

The AVBM offers the following advantages: (a) It is suitable for detection of small defects and progressive damage of large structures and (b) effective for both global and localized monitoring. The challenges are (a) sensitivity to environmental factors (e.g., temperature, humidity) that degrades the output and (b) requires a robust baseline model for comparison.

The active vibration-based method is particularly useful for SHM of composite materials in aerospace (monitoring composite structures in aircrafts), wind turbine blades, and civil infrastructure (bridges and composite pipelines), where real-time damage detection and structural integrity assessment are critical.

The AVBM can be categorized as either "model-based" or not based on models. The "model-based' approach involves analyzing models of structures, and utilizing characteristic (modeling) parameters to locate defects. In contrast, the second mode of approach enables damage detection from changes in vibration characteristics, and do not depend on structural models.

Model-based AVBMs are more commonly used in real-time SHM applications, and comprises of three categories. The first, referred to as "forward problems" category, involves calibration of parameters (used in modeling) for different, historical damage scenarios. Defects are estimated by comparing the measured and the predicted values of these parameters. The primary challenge of this approach lies in acquiring sufficient and accurately-measured characteristic parameters for all structural scenarios. This becomes especially difficult for larger structures where experimental measurements may be impractical, and computational techniques such as the finite element method (FEM) are more viable. The next category of methods involves defining "criteria" or "indicators" to analyze variations in model parameters for identification of structural damages (flaws). The common natural frequency criteria (CNFC) include the Cawley–Adams criterion and the damage location assurance criterion (DLAC). For locating multiple flaw sites, the multiple damage location assurance criterion (MDLAC) is used, which is an extension of the DLAC. The criteria based on frequency response functions (FRFs) include the frequency domain assurance criterion (FDAC), the global amplitude correlation (GAC) function, the frequency response assurance criterion (FRAC), and the global shape correlation (GSC) function. The modal assurance criterion (MAC) quantifies the correlation between the analytically-obtained and the measured mode shapes as a scalar parameter that ranges from zero to unity. Enhanced versions of MAC include the partial MAC (PMAC) and the co-ordinate MAC (COMAC). Additionally, variations in other model parameters, such as dynamic flexibility and the mode shape curvature, can also be calculated and used as indicators. The third category is based on modifications of the model, wherein differences between the original and the modified (structural) models are used to extract information on damage. This approach utilizes measured data, analytical models, and the equations of structural dynamics, and is mathematically formulated as a constrained optimization problem. It is worthwhile to mention that, unlike the active Lamb wave or impedance methods, the vibration frequencies used in AVBM are typically much lower [10, 52, 53].

9.5.4 Acoustic Emission "Passive" Methods for SHM

The acoustic emission (AE) method is an example of non-destructive technique that is widely used for structural health monitoring. It detects and analyzes transient (i.e., time-dependent), elastic waves generated by the rapid rate of strain energy-release from localized damage episodes (such as crack propagation, delamination, or fiber breakage) within the material. The AE technique proves effective for detecting and monitoring damage (cracks) progression in composites, and provides valuable insights into their structural integrity [46]. The operational principle of the AE is now discussed: Sensors placed immediately beneath the surface of a composite structure capture the acoustic signals produced by damage events (such as crack propagation). The amplitude, frequency, and location of these signals are thereafter analyzed to identify the type, severity, and position of the damage. Signal-processing techniques and algorithms are employed to differentiate between "noise" and actual damage-related emissions. To summarize, acoustic waves are generated at the source and then propagate to the sensors. The sensors, then, capture these waves and convert them into electrical signals, which are processed and analyzed, in order to assess the structural integrity. Since AE testing relies solely on sensors to passively detect signals, it is classified as a passive SHM technique. Primary AEs occur during events that lead to damage within structures; in contrast, secondary AEs are those triggered by external sources (e.g., impacts). Even in cases when a structure experiences damages at small-length scales, the AE phenomena enables defect-prediction and real-time monitoring.

The triangulation method is a widely used approach for crack location that determines the location of a crack (or damage) in the acoustic emission technique. It relies on measuring the time-differences-of-arrival (TDOA) of acoustic waves at multiple sensors (at least three) placed on the surface of the monitored structure. The basis of the triangularization method is as follows: When a crack or damage occurs, it generates acoustic waves that travel through the (composite) structure. A network of AE sensors, placed at known positions, detects these waves. The arrival times (TOA) of the waves is recorded at each sensor. The differences in the wave's arrival times (or TDOA) between the sensor pairs are used to calculate the source's location. By combining the TDOA data from at least three sensors, the position of the crack or damage is pinpointed using the geometric principles of "triangulation". The principal factors determining the effectiveness of the triangulation method are (a) wave speed (speed of the acoustic wave in the material must be known, or estimated, for accurate calculations), (b) sensor arrangements (accurate placement of sensors is critical for effective triangulation and minimizing errors), and (c) signal noise (high signal-to-noise, or SINR ratios improve accuracy in determining the source location). The advantages of this method are (a) enabling precise localization of cracks or (structural damages), (b) effectiveness in real-time monitoring of large (composite) structures, and (c) minimal (invasive) disruption to the structure during operation. To summarize, the triangulation method is particularly useful in SHM for composite materials, where detecting and locating internal damage such as delamination or fiber breakage is crucial [10, 54].

9.5.5 Strain-Based Methods for SHM

The strain-based method is a widely used approach in structural health monitoring (SHM) of composites, focusing on the measurement and analysis of strain to detect and evaluate damage. This method is a passive and effective SHM technique, as damages in (composite) structures modify the strain fields locally by altering the stress field (load contours), even under normal operational loads. This method relies on the principle that structural damage, such as cracks, delamination, or debonding, alters the strain distribution within the material. Sensors (e.g., strain gauges, FBG sensors or piezoelectric sensors) are attached to, or embedded in, the composite structure to monitor strain. These sensors detect anomalies in strain, deviations or gradients, which indicate the presence of deformation or fracture. The recorded data is analyzed to locate and quantify the structural changes, enabling real-time detection and assessment of damage. In addition to resistance strain gauges, FOSs are also used to determine distributed strains.

Strain gauges are widely used for measuring strains in (composite) materials and structures. They operate on the following physical principle: the electrical resistance of a material changes when subjected to geometrical deformations. Strain gauges are typically thin, metallic or semiconductor elements bonded to the surface of a structure. When the structure undergoes strain, the strain gauge stretches or compresses, whereby its length and/or its cross-sectional area changes, and this alters its electrical resistance. Typically, a Wheatstone bridge circuit that measures the change in resistance is calibrated to determine the strain. The types in use are (a) foil strain gauges, offering high sensitivity and durability, (b) semiconductor strain gauges, which offer still higher ranges of sensitivity but less stability over time, and (c) wire strain gauges, used in high-temperature and/or dynamic environments.

Piezoelectric sensors utilize the piezoelectric effect to measure physical parameters (e.g., pressure, strain, acceleration, or force) pertaining to safe structural requirements. They are widely used in SHM practices for their high sensitivity, compactness, and ability to operate in harsh environments. Their working principle is as follows: Certain materials (e.g., quartz or piezoelectric ceramics) generate an EMF when subjected to mechanical/thermal stress. The electrical potential difference is proportional to the applied stress (magnitude), and allows measurement of dynamic forces or vibrations. The (electrical) signals are processed and analyzed to assess structural condition of composites. The advantages of piezoelectric sensors are (a) high sensitivity to small deformations and dynamic changes, (b) lightweight and compact dimensions, which are suitable for embedded applications, (c) capacity to operate in extreme temperatures and environments, and (d) wide range of frequency response for detection of both low- and high-frequency signals.

FBG sensors, as advanced optical sensors, are also used for strain-based SHM monitoring. They are highly valued in structural health monitoring due to their (a) high precision, (b) multiplexing capabilities, and (c) immunity to electromagnetic interference. Their working principle is as follows: FBGs are fabricated by inscribing a periodic variation in the refractive index along the core of an optical fiber. When light rays travels through the fiber, the fiber grating reflects a specific Bragg wavelength, while transmitting the rest. Changes in strain, load, or temperature alter the grating's periodicity and refractive index, producing a shift in the reflected wavelength (λ_B). This shift is measured for determining the parameters connected with SHM. The advantages offered by FBG sensors are (a) high sensitivity and accuracy, (b) capability of multiplexing multiple FBGs on a single fiber for distributed sensing, (c) lightweight and compact, (d) resistant to harsh environments, and (e) immunity to electromagnetic interference [55].

In practical applications, there are two approaches by which the strain-based method can be implemented. The first approach involves measurement of strain distribution of an undamaged (composite) structure beforehand; this serves to establish a baseline. Thereafter, distortion (or damage) is identified when the current strain measurements deviate from this baseline. The second approach involves modelling and simulations, whereby an analytical (structural) model is developed and analyzed to obtain the strain field under different loading conditions. By comparing the analytical data with the actual strain measurements—either directly or using a predefined criterion—the structural integrity is assessed. The primary challenge with this (second) approach is developing sufficiently accurate models for complex, real-world structures [10, 56].

9.5.6 The Comparative Vacuum Monitoring (CVM) "Passive" Method

The CVM method is a non-destructive, SHM technique used for detection and monitoring of damage (such as cracks, delamination, or voids) in composite materials. It is based on comparison of the pressure levels in a network of small, interconnected vacuum channels embedded or bonded to the surface of the structure. Designed and developed by Structural Monitoring System Ltd. (SMSL), the CVM system consists of three main components: a stable low-vacuum source, a fluid flow meter, and a CVM sensor. The CVM sensor is directly attached to the surface of the structure (being monitored), and consists of a network of elongated, narrow galleries that alternate between low-vacuum and atmospheric conditions [57]. The stable vacuum source, monitored by the flow meter and sensors, provides a reference pressure for the vacuum galleries. When the structure is intact, the galleries remain sealed, resulting in zero air leakage or pressure changes. When However, the galleries are breached by a structural flaw, atmospheric air flows through the damaged passages into the vacuum galleries, and causes pressure rise [58, 59]. The rate of increase in pressure can be calibrated to serve as an indicator of the damage size. The comparison of pressure levels (in damaged and undamaged sections) allows precise identification and localization of the damage.

The thickness of the gallery walls influences the sensitivity of the CVM sensor [60]. The CVM technique is particularly well-suited for detecting corrosion in metals or surface cracks, as the exposed structural surfaces form part of galleries. Remote inspection of debonding, delamination, or crack growth in composites can be effectively performed with embedded CVM sensors [10, 11].

9.6 Conclusions and Future Scope

The monitoring of structural health for composites has become an inevitable aspect towards ensuring the reliability and longevity of composite structures. Based on published literature, this chapter presents an outline of the application potential of composite materials in various fields of technology. The chapter highlights their advantages, inherent limitations and implementation challenges, and the necessity of SHM for maintaining their structural integrity. A review of different sensor technologies has been presented, along with an in-depth discussion of the key, state-of-the-art SHM methodologies such as the Lamb wave method, active vibration-based method, electromechanical impedance method, acoustic emission methods, strain-based techniques, and the comparative vacuum monitoring (CVM) method.

In spite of significant advancements in SHM technologies and practice-protocols, several challenges persist in terms of real-time monitoring, sensor integration, data processing, and cost-effectiveness. Future R&D should focus on enhancement of sensor performance under adverse environmental conditions, and development of multi-functional sensor networks. R&D efforts should also leverage artificial intelligence and machine learning for automated fault-diagnosis, damage-detection and predictive maintenance. In addition, advancements in wireless sensor networks and energy-harvesting techniques will serve to improve the feasibility of SHM for large-scale composite structures. By addressing these challenges, SHM systems can gradually and seamlessly evolve into highly efficient, self-sustaining solutions, and shall ensure safer and more cost-effective use of composite materials in diverse engineering applications.

References

1. Gibson, R.F., A review of recent research on mechanics of multifunctional composite materials and structures. *Compos. Struct.*, 92, 12, 2793–2810, 2010.
2. Mylsamy, B., Aruchamy, K., Shanmugam, S.K.M., Palanisamy, S., Ayrılmis, N., Improving performance of composites: Natural and synthetic fibre hybridisation techniques in composite materials–A Review. *Mater. Chem. Phys.*, 130439, 2025.
3. De, B., Bera, M., Bhattacharjee, D., Ray, B.C., Mukherjee, S., A comprehensive review on fiber-reinforced polymer composites: Raw materials to applications, recycling, and waste management. *Prog. Mater Sci.*, 146, 101326, 2024.
4. Asfaw, N.T., Absi, R., Labouda, B.A., Abbassi, I.E., Assessment of the thermal and mechanical properties of bio-based composite materials for thermal insulation: A review. *J. Build. Eng.*, 97, 110605, 2024.
5. Yadav, R., Singh, M., Shekhawat, D., Lee, S.Y., Park, S.J., The role of fillers to enhance the mechanical, thermal, and wear characteristics of polymer composite materials: A review. *Compos. Part A: Appl. Sci. Manuf.*, 175, 107775, 2023.
6. Singh, B., Gupta, M., Verma, A., Mechanical behaviour of particulate hybrid composite laminates as potential building materials. *Constr. Build. Mater.*, 9, 1, 39–44, 1995.

7. Ribeiro Filho, S.L.M., Garcia, C.T., Donadon, M.V., Scarpa, F., Panzera, T.H., The impact behaviour of hybrid fibre-particle composites based on a full factorial design. *Mater. Today Commun.*, 31, 103459, 2022.
8. Zhang, Q., Huang, J.Q., Qian, W.Z., Zhang, Y.Y., Wei, F., The road for nanomaterials industry: A review of carbon nanotube production, post-treatment, and bulk applications for composites and energy storage. *Small*, 9, 8, 1237–1265, 2013.
9. Selvaraju, S. and Ilaiyavel, S., Applications of composites in marine industry. *J. Eng. Res. Stud.*, 2, 2, 89–91, 2011.
10. Hu, N. (Ed.), *Composites and their applications*, IntechOpen, Rijeka, Croatia, 2012, https://doi.org/10.5772/2318.
11. Elhajjar, R., La Saponara, V., Muliana, A. (Eds.), *Composite materials: Analysis and design*, CRC Press, Taylor & Francis Group, Boca Raton, FL, 2014, https://doi.org/10.1007/978-3-319-08094-8.
12. Lau, K.T., Zhou, L.M., Tse, P.C., Yuan, L.B., Applications of composites, optical fibre sensors and smart composites for concrete rehabilitation: an overview. *Appl. Compos. Mater.*, 9, 4, 221–247, 2002.
13. Islam, M.H., Afroj, S., Uddin, M.A., Andreeva, D.V., Novoselov, K.S., Karim, N., Graphene and CNT-based smart fiber-reinforced composites: A review. *Adv. Funct. Mater.*, 32, 40, 2205723, 2022.
14. Olabi, A.G. and Grunwald, A., Design and application of magnetostrictive materials. *Mater. Des.*, 29, 2, 469–483, 2008.
15. Hristoforou, E. and Ktena, A., Magnetostriction and magnetostrictive materials for sensing applications. *J. Magn. Magn. Mater.*, 316, 2, 372–378, 2007.
16. Umar, K., Yaqoob, A.A., Ibrahim, M.N.M., Parveen, T., Safian, M.T.U., Environmental applications of smart polymer composites, in: *Smart Polymer Nanocomposites*, pp. 295–312, Woodhead Publishing, Cambridge, United Kingdom, 2021.
17. Lau, K.T., Fibre-optic sensors and smart composites for concrete applications. *Mag. Concr. Res.*, 55, 1, 19–34, 2003.
18. de Medeiros, R., Lopes, H.M., Guedes, R.M., Vaz, M.A., Vandepitte, D., Tita, V., A new methodology for structural health monitoring applications. *Procedia Eng.*, 114, 54–61, 2015.
19. Farrar, C.R. and Worden, K., An introduction to structural health monitoring. *Philos. Trans. R. Soc. A: Math. Phys. Eng. Sci.*, 365, 1851, 303–315, 2007.
20. Tian, Y., Chen, C., Sagoe-Crentsil, K., Zhang, J., Duan, W., Intelligent robotic systems for structural health monitoring: Applications and future trends. *Autom. Constr.*, 139, 104273, 2022.
21. Awad, Z.K., Aravinthan, T., Zhuge, Y., Gonzalez, F., A review of optimization techniques used in the design of fibre composite structures for civil engineering applications. *Mater. Des.*, 33, 534–544, 2012.
22. Montalvao, D., Maia, N.M.M., Ribeiro, A.M.R., A review of vibration-based structural health monitoring with special emphasis on composite materials. *Shock Vib. Dig.*, 38, 4, 295–324, 2006.

23. Kossakowski, P.G. and Wciślik, W., Fiber-reinforced polymer composites in the construction of bridges: Opportunities, problems and challenges. *Fibers*, 10, 4, 37, 2022.
24. Qing, X., Kumar, A., Zhang, C., Gonzalez, I.F., Guo, G., Chang, F.K., A hybrid piezoelectric/fiber optic diagnostic system for structural health monitoring. *Smart Mater. Struct.*, 14, 3, S98, 2005.
25. Qing, X.P., Chan, H.L., Beard, S.J., Kumar, A., An active diagnostic system for structural health monitoring of rocket engines. *J. Intell. Mater. Syst. Struct.*, 17, 7, 619–628, 2006.
26. Ihn, J.B. and Chang, F.K., Pitch-catch active sensing methods in structural health monitoring for aircraft structures. *Struct. Health Monit.*, 7, 1, 5–19, 2008.
27. Hassani, S., Mousavi, M., Gandomi, A.H., Structural health monitoring in composite structures: A comprehensive review. *Sensors*, 22, 1, 153, 2021.
28. Ricci, F., Monaco, E., Boffa, N.D., Maio, L., Memmolo, V., Guided waves for structural health monitoring in composites: A review and implementation strategies. *Prog. Aerosp. Sci.*, 129, 100790, 2022.
29. Li, H.N., Li, D.S., Song, G.B., Recent applications of fiber optic sensors to health monitoring in civil engineering. *Eng. Struct.*, 26, 11, 1647–1657, 2004.
30. Kahandawa, G.C., Epaarachchi, J., Wang, H., Lau, K.T., Use of FBG sensors for SHM in aerospace structures. *Photonic Sensors*, 2, 203–214, 2012.
31. Zhou, Z. and Ou, J., Development of FBG sensors for structural health monitoring in civil infrastructures, in: *Sensing issues in civil structural health monitoring*, pp. 197–207, Springer, Dordrecht, Netherlands, 2005.
32. Lau, K.T., Structural health monitoring for smart composites using embedded FBG sensor technology. *Mater. Sci. Technol.*, 30, 13, 1642–1654, 2014.
33. Čápová, K., Velebil, L., Včelák, J., Laboratory and *in-situ* testing of integrated FBG sensors for SHM for concrete and timber structures. *Sensors*, 20, 6, 1661, 2020.
34. Torres, B., Payá-Zaforteza, I., Calderón, P.A., Adam, J.M., Analysis of the strain transfer in a new FBG sensor for structural health monitoring. *Eng. Struct.*, 33, 2, 539–548, 2011.
35. Dos Reis, J., Oliveira Costa, C., Sá da Costa, J., Strain gauges debonding fault detection for structural health monitoring. *Struct. Control Health Monit.*, 25, 12, e2264, 2018.
36. Webb, G.T., Vardanega, P.J., Middleton, C.R., Categories of SHM deployments: Technologies and capabilities. *J. Bridge Eng.*, 20, 11, 04014118, 2015.
37. Jo, H., Park, J.W., Spencer, B.F., Jung, H.J., Develoment of high-sensitivity wireless strain sensor for structural health monitoring. *Smart Struct. Syst.*, 11, 5, 477–496, 2013.
38. Lin, B. and Giurgiutiu, V., Modeling and testing of PZT and PVDF piezoelectric wafer active sensors. *Smart Mater. Struct.*, 15, 4, 1085–1093, 2006.
39. Boller, C., Next generation structural health monitoring and its integration into aircraft design. *Int. J. Syst. Sci.*, 31, 11, 1333–1349, 2000.

40. Raghavan, A. and Cesnik, C.E.S., Review of Guided-wave Structural Health Monitoring. *Shock Vib. Dig.*, 39, 2, 91–114, 2007.
41. Sodano, H.A., Development of an automated eddy current structural health monitoring technique with an extended sensing region for corrosion detection. *Struct. Health Monit.*, 6, 2, 111–119, 2007.
42. Zilberstein, V., Walrath, K., Grundy, D., Schlicker, D., Goldfine, N., Abramovici, E., Yentzer, T., MWM eddy-current arrays for crack initiation and growth monitoring. *Int. J. Fatigue*, 25, 9–11, 1147–1155, 2003.
43. Varadan, V.K. and Varadan, V.V., Microsensors, microelectromechanical systems (MEMS), and electronics for smart structures and systems. *Smart Mater. Struct.*, 9, 6, 953–972, 2000.
44. Yuan, S., Liang, D., Shi, L., Zhao, X., Wu, J., Li, G., Qiu, L., Recent progress on distributed structural health monitoring research at NUAA. *J. Intell. Mater. Syst. Struct.*, 19, 3, 373–386, 2008.
45. Qing, X.P., Beard, S.J., Kumar, A., Chan, H.L., Ikegami, R., Advances in the development of built-in diagnostic system for filament wound composite structures. *Compos. Sci. Technol.*, 66, 11–12, 1694–1702, 2006.
46. Lemistre, M.B. and Balageas, D.L., A hybrid electromagnetic acousto-ultrasonic method for SHM of carbon/epoxy structures. *Struct. Health Monit.*, 2, 2, 153–160, 2003.
47. Giurgiutiu, V., Structural health monitoring with piezoelectric wafer active sensors–predictive modeling and simulation. *INCAS Bull.*, 2, 3, 31–44, 2010.
48. Su, Z. and Ye, L., Selective generation of Lamb wave modes and their propagation characteristics in defective composite laminates. *Proc. Inst. Mech. Eng., Part L: J. Mater.: Des. Appl.*, 218, 2, 95–110, 2004.
49. Quek, S.T., Tua, P.S., Jin, J., Comparison of plain piezoceramics and interdigital transducer for crack detection in plates. *J. Intell. Mater. Syst. Struct.*, 18, 9, 949–961, 2007.
50. Su, Z., Ye, L., Bu, X., A damage identification technique for CF/EP composite laminates using distributed piezoelectric transducers. *Compos. Struct.*, 57, 1–4, 465–471, 2002.
51. Park, G., Sohn, H., Farrar, C.R., Inman, D.J., Overview of piezoelectric impedance-based health monitoring and path forward. *Shock Vib. Dig.*, 35, 6, 451–464, 2003.
52. Farrar, C.R., Doebling, S.W., Nix, D.A., Vibration–based structural damage identification. *Philos. Trans. R. Soc. Lond. Ser. A: Math. Phys. Eng. Sci.*, 359, 1778, 131–149, 2001.
53. Messina, A., Williams, E.J., Contursi, T., Structural damage detection by a sensitivity and statistical-based method. *J. Sound Vib.*, 216, 5, 791–808, 1998.
54. Nair, A. and Cai, C., S. Acoustic emission monitoring of bridges: Review and case studies. *Eng. Struct.*, 32, 6, 1704–1714, 2010.
55. Jeong, H., Analysis of plate wave propagation in anisotropic laminates using a wavelet transform. *NDT & E Int.*, 34, 3, 185–190, 2001.

56. Silva-Muñoz, R.A. and Lopez-Anido, R.A., Structural health monitoring of marine composite structural joints using embedded fiber Bragg grating strain sensors. *Compos. Struct.*, 89, 2, 224–234, 2009.
57. Kurnyta, A., Dragan, K., Dziendzikowski, M., Assessment of sensor technologies for aircraft SHM systems. *Fatigue Aircr. Struct.*, 2013, 5, 53–59, 2013. Wishaw, M. and Barton, D.P., Comparative vacuum monitoring: A new method of *in-situ* real-time crack detection and monitoring, in: *Proceedings of the 10th Asia-Pacific Conference on Nondestructive Testing*, Brisbane, Australia, 18-21, September, 2001.
58. Perterer, M., Friemel, M., Baier, H., Characterization of the Mechanical Influence of Comparative Vacuum Measurement (CVM) Sensors in the Context of Structural Health Monitoring (SHM) Systems, in: *6th European Workshop on Structural Health Monitoring*, 1–8, July 2012.
59. Stehmeier, H. and Speckmann, H., Comparative vacuum monitoring (CVM), in: *Proceedings of the 2nd European Workshop on Structural Health Monitoring*, Munich, Germany, 7–9, July 2004.
60. Roach, D., Real time crack detection using mountable comparative vacuum monitoring sensors. *Smart Struct. Syst.*, 5, 4, 317–328, 2009.

Index

0-0 composites, 78
0-2 composites, 78
0-3 composites, 78–79

3.5%-NaCl solution, 103

Additive manufacturing, 123, 191, 193–194, 197, 210
Aerodynamic force, 117
Aluminum nitride (AlN), 88–89
Aspect ratio, 204, 206–210
ASTM E384, 103, 104
Austenite phase, 21–22, 24

BCZT (Barium Calcium Zirconate Titanate), 89
Biocompatibility, 127
Biodegradability, 127
Bio-medical applications, 89
Biomedical engineering, 121
BNT (Bismuth Sodium Titanate), 88
Boundary conditions, 204, 206–210

Carbon fiber, 16
Carbon fiber-reinforced polymer (CFRPs), 115, 125
Cellular polypropylene (PP), 86–87
Cellulose nanocrystals (CNC), 180
Ceramic-based nanocomposites, 87–89
Ceramic-rich, 191, 197–199, 206
CFFF plate, 205, 207
Clamped plate, 204
Composite material, 189–191, 194, 197
Composites
  problems and challenges, 220
  smart composites, 218–219
  traditional composites, 216–217
Composites for actuation, 59, 60
Conductivity, 117
Constraints, 179–180
Contact poling, 82–83
Corona poling, 82
Corrosion resistance, 120, 126
Current density, 109

Delamination, 189, 191–192
Dielectric constant (ε), 77–78
Differential scanning calorimetry, 34
Direct piezoelectric effect, 73, 75–76
Displacement field, 200
Dynamic behavior, 191, 197

Electrical loss (tan δ), 77
Electrochemical, 108
Electro-mechanical coupling factor (k), 74–75
Electrospinning, 123
Energy harvesters, 90
Equal channel angular pressing, 45

Face-centered cubic (FCC), 24
FEM, 189, 196–197, 210–211
FGM/functionally graded, 189–198, 203–204, 206, 208–211
Free vibration, 189, 197, 204, 210
FSDT, 200, 203

Functionality, 50, 59, 65
Fundamental frequency, 206, 208–210
Fuselages, 120

Glass fiber-reinforced polymers (GFRPs), 115, 126

Hafnium/zirconium alloys, 45
Hall–Petch relation, 108
Hamilton's principle, 202, 204
High wear applications, 169–170, 175, 177, 181
Hydrogel, 117, 124
Hypersonic, 121

Improved mechanical properties, 172, 178
Isogeometric analysis (IGA), 196
Isopropanol, 101
Izod test, 55

Kevlar, 16
Kinetic energy, 202–203
KNN (Potassium Sodium Niobate), 88

Length-to-thickness ratio, 207
Lithium niobate, 89

Macroscopically, 116
Magnetostrictive, 49, 51
Martensite phase, 23
Mechanical quality factor (Qm), 77
MEMS actuators, 37
Mesh, 206
Metal matrix composite (MMC, 100
Metal-rich, 191, 197–198, 206
Microhardness, 108
Microstructural evolution, 170, 183
Mid-plan, 200–201
Mode shapes, 206–207

Nanocomposite electrical generators (NEG), 93–94
Nanotechnology-driven production, 116
Natural fiber, 3, 10, 12
Natural frequency, 189, 204
Ni–50% ZrO_2–TiO_2, 105
Nickel plating, 100
Nitinol (NiTi), 23
Node, 201
Non-destructive testing, 55
Non-dimensional, 204–207
Nucleation, 100

Orthorhombic, 106, 107
Overpotential, 107

Payload capacity, 119
Piezoelectric, 49, 51, 59, 60, 61, 117, 123
Piezoelectric nanocomposites, 78–79
Piezoelectric nano-generator (PENG), 90–93
Piezoelectric strain constant (d), 75–76
Piezoelectric tactile sensors, 89–90
Piezoelectric voltage constant (g), 76
Piezoelectricity, 72–73, 76–77
Poisson's ratio, 197–198, 203
Poling process, 81–82
Polyamide 11 (Nylon 11), 86
Polyethylene glycol, 103
Polymer matrix composites (PMCs), 115, 125
Polymer-based nanocomposites, 79–81
Power law index, 199–210
Pseudo elasticity, 22
Pulsed electrodeposition, 100
PVDF (Polyvinylidene Fluoride), 85
PVDF-TrFE, 86
PZT (Lead Zirconate Titanate), 87–88

Quadrilateral element, 201

Real-world, 49, 64, 67
Rectangular pulse, 107
Research pathways, 180
Reverse piezoelectric effect, 73, 74
Rotor blade, 120

SCSC plate, 204
Self-healing, 49, 50, 51, 53, 57, 60, 61, 62, 63, 66, 67, 68
Self-healing polymer, 115
Sensors
  eddy current sensors, 219, 221–222, 225–226, 228
  fiber optic sensors, 219, 222–224, 239
  MEMS sensor, 220, 222, 226–228
  piezoelectric sensors, 220, 222, 225, 230, 232–234, 238–239
  resistance strain gauges, 219, 222, 224–225, 227, 238
Severe plastic deformation, 45
Shape function, 201–202
Shape memory foams, 46
Shape memory polymers (SMPs), 115, 127, 128
Shear, 200–201, 203
Shear correction factor, 203
Sigmoid law, 197–199
Simply supported plate (SSSS), 204
Smart composites, 49, 59, 61, 62, 63, 64, 65, 66, 67
Smart functional, 55, 58, 60, 64, 65, 65, 67, 69
Smart functional composites, 169–170, 172, 177, 179, 182
Sol-gel, 101
Strain energy, 202
Stress-induced martensitic, 29
Structural health monitoring, 219
  active structural health monitoring
    electromechanical impedance method, 220, 229, 232–234, 236, 241
    lamb wave method/technique, 220, 225, 229–232, 236, 240
    vibration-based monitoring, 220, 225, 227, 234–236, 241
  benefits and challenges, 221–222
  passive structural health monitoring
    acoustic emission method, 220, 225, 227, 229, 236–238, 241
    comparative vacuum monitoring (CVM), 220, 228–229, 240–241
    strain-based monitoring method, 220, 229, 238–239, 241
Structural smart composites, 59
Structural stiffness, 207
Super elastic plateau, 29
Super paramagnetic, 129
Supersonic passenger airliner, 120
Surface coatings, 100

Tensile testing, 55
Thermal stress, 197
Thermomechanical hysteresis, 28
Thermoplastic, 5, 6, 9, 10, 11, 14
Thermoset, 5, 9, 10, 11
TiO_2 and ZrO_2, 100
Transformation Temperatures, 28
Triangular pulse, 108
Triangular wave, 101
Tribological coating, 169–170, 172–173, 177, 179, 182
Tribological performance, 169, 179, 183
Turbofan engine blade, 120

UV radiation, 120

Vickers hardness test, 54
Viscoelasticity, 117
Voigt law, 199
Volume fraction, 198–199

Wear resistance, 169, 172–175, 177–179, 181